QUANTIFYING THE AGRI-FOOD SUPPLY CHAIN

Wageningen UR Frontis Series

VOLUME 15

Series editor:

R.J. Bogers

Frontis – Wageningen International Nucleus for Strategic Expertise, Wageningen University and Research Centre, Wageningen, The Netherlands

Online version at http://www.wur.nl/frontis

Cover figure based on: Lazzarini, S.G., Chaddad, F.R. and M.L. Cook, 2001. Integrating supply chain and network analyses: the study of netchains. *Journal on Chain and Network Science*, 1(1), 7-22.

The titles pub lished in this series are listed at the end of this volume.

QUANTIFYING THE AGRI-FOOD SUPPLY CHAIN

Edited by

CHRISTIEN J.M. ONDERSTEIJN

Business Economics Group, Wageningen University,
The Netherlands

JO H.M. WIJNANDS

Business Economics Group, Wageningen University,
The Netherlands

RUUD B.M. HUIRNE

Animal Science Group, Wageningen University,
The Netherlands

and

OLAF VAN KOOTEN

Horticultural Production Chains Group, Wageningen University,
The Netherlands

A C.I.P. Catalogue record for this book is available from the Library of Congress.

ISBN-10 1-4020-4692-8 (HB)
ISBN-13 978-1-4020-4692-6 (HB)
ISBN-10 1-4020-4693-6 (PB)
ISBN-13 978-1-4020-4693-3 (PB)

Published by Springer,
P.O. Box 17, 3300 AA Dordrecht, The Netherlands.

www.springer.com

Printed on acid-free paper

Printed in the Netherlands.

CONTENTS

Modelling agri-food chains

The value of information in agri-food chains

Supply chain organization and chain performance

PREFACE

Due to globalization and internationalization of production, the arena of competition and competitive advantage is moving from individual firms operating on spot markets towards supply chains and networks. Therefore, coordination mechanisms between firms within the chain become more important. It took decades before research explicitly started to focus on (the reason behind) vertical relationships, production chains and networks. This resulted in a large body of research on transaction costs. While these costs have been looked at extensively, the costs of producing in chains and the chain processes have received relatively little attention. Topics like costs, efficiency, risk and investment analysis have hardly received any empirical attention within chain and network research. Nonetheless, these performance measures are of vital importance for continuity of individual companies, chains and networks.

Even though many people attempt to, and actually do, quantify supply-chain performance, risk and investment behaviour, it has mostly been done more or less *ad hoc*. To develop a more coherent view on this matter, a workshop with experts in the field was held in October 2004 at Wageningen, The Netherlands. The goal of the workshop was to develop a framework for quantification of costs, benefits, efficiency and risk in agri-food chains. The challenge was to compare agri-food chains on their effectiveness and to bridge the gap between management science and the technical sciences. Thirty scientists participated in the workshop, which was organized by Frontis – Wageningen International Nucleus for Strategic expertise, the Business Economics Group of the Department of Social Sciences, and the Horticultural Production Chains Group of the Department of Plant Sciences, all part of Wageningen University and Research Centre.

FRAMEWORK OF THE BOOK

The workshop was organized in five sessions: 'Sharing costs, benefits and risk in agri-food chains', 'Measuring performance in agri-food chains', 'Modelling agri-food chains', 'The value of information in agrifood chains' and 'Supply chain organization and chain performance'. Within each session, presentations included different perspectives and approaches. The state of the art, areas for new research and areas where knowledge from other scientific disciplines is needed, were addressed. This book follows the same outline: five sections organized around the papers presented at the workshop.

ACKNOWLEDGEMENTS

We would like to thank the authors for their contributions, Petra van Boetzelaer for making the organizational arrangements for the workshop, Christien Ondersteijn and

Jo Wijnands for editing the manuscripts and Jos Smelik for facilitating the lay-out process.

Robert J. Bogers
Ruud B.M. Huirne
Olaf van Kooten

Wageningen, September 2005

INTRODUCTION

CHAPTER 1

QUANTIFYING THE AGRI-FOOD SUPPLY CHAIN

Overview and new research directions

JO H.M. WIJNANDS AND CHRISTIEN J.M. ONDERSTEIJN

Business Economics Group, Wageningen University, Hollandseweg 1, 6706 KN Wageningen, The Netherlands. E-mail: jo.wijnands@wur.nl

Abstract. The Frontis workshop 'Quantifying the agri-food supply chain' aimed at discussing the possibilities and limitations of quantifying performance, risks and investments in the agri-food chain and at bringing people from international institutes together. Their contributions are organized around five key issues in the agri-food chain: concepts of measuring performance; empirical research in measuring costs, benefits and risk; modelling; value of information; and governance and performance.

Papers with a wide variety of approaches from different economic disciplines have been demonstrated to be useful in analysing the supply chain. However, understanding the complex system of agri-food chains requires more investments in retrieving empirical data for testing propositions and developing appropriate models. The identified research gaps and discussion points were shared among an international forum of researchers. International cooperation among researchers will enhance progress in this research field. The workshop was highly valued in this respect.
Keywords: agribusiness; food production chains; performance measurements; risk and uncertainty; chain governance

INTRODUCTION

The chapters in this volume were first presented at a Frontis workshop on 'Quantifying the agrifood supply chain', held at Wageningen, The Netherlands, 22-24 October 2004. The overall purpose of this workshop was to discuss the possibilities and limitations of quantifying performance, risks and investments in the agri-food chain and to bring people from international institutes together. Agri-food chains entered a new era in which customer orientation and social responsibility are the main driving forces. Globalization of supply chains complicates the chain governance. In order to develop a research agenda that meets the challenges facing industry and policymakers, invited experts from around the world convened to review the state of the art.

The papers in this volume were written on invitation and are organized around five key issues in the agri-food chain:

C.J.M. Ondersteijn et al. (eds.), Quantifying the agri-food supply chain, 3-12.

1. Concepts of measuring performance.
2. Empirical research in measuring costs, benefits and risk.
3. Modelling.
4. Value of information.
5. Governance and performance.

In the following overview we provide the highlights of each key issue and important insights obtained from the presentations and discussions. In the conclusions we define research areas for future work.

MEASURING PERFORMANCE IN AGRI-FOOD CHAINS

At firm level a large number of performance indicators are available; see for instance the balanced scorecard of Kaplan and Norton (1992). Performance indicators at chain level are still being developed. First of all, performance indicators on chain level have to deal with the governance structure of the chain. The objectives of chain partners are not necessarily in line with optimal performance of the total chain. Secondly, the relevance of information differs on each level, even if the information is of high importance for the overall chain performance. Thirdly, the strategic value of some of the information inhibits a free exchange between chain partners.

Many researchers and practitioners are working on the enhancement of supply-chain collaboration in order to improve performance of the individual supply-chain members and supply-chain performance as a whole. The first group of papers provides an overview of approaches of chain performance measures.

Van der Vorst presents a framework for the development of innovative food supply-chain networks and discusses the implications for performance measurement systems. Performance measurement fulfils a crucial role in the development of supply chains and networks as it can direct the design and management of the chain towards the required performance. It is the key instrument to discuss and evaluate the effectiveness of (potential) chain partnerships. He defines supply chains and supply-chain management. Furthermore, he distinguishes Key Performance Indicators (KPIs) on chain, organization and process level. Lack of transparency and cooperation are the main bottlenecks in measuring chain performances. Many indicators focused on local operational use and are lacking standard definitions. He also points out that academics are interested in bringing definitions, adaptation and validation under one heading. This conflicts in many cases with the managers' view that providing useful information quickly is a good enough measure.

Principles for accounting in supply chains are the topic of Bremmers's paper. The availability of (normative) accounting standards is a prerequisite for effective and efficient accounting in supply chains. He proposes three accounting standards for cooperative supply chains:

- Reciprocity in information access: those that deliver information to the system should be able to retrieve an equivalent amount.
- Equivalent cash flows: provision of assets (supply-chain investments) should be matched with an equivalent amount of assets (cash flows) in return.

- Matching risks and returns: the bigger the opportunities for individual firms, the bigger the contribution should be in risk sharing and risk management.

These three rules of reciprocity cause a transparent supply-chain policy and performance measurement. Transparency will replace (or at least will have to supplement) *trust* as a measure for relational quality. These three standards are not meant to be exhaustive; other standards should still be developed. Moreover, he argues that technological innovation, such as electronic chain-wide reporting, is beneficial for transparency, decision making and control in/of supply chains, and will reduce the administrative costs of (supply-chain) accounting systems at the same time.

The topic of Bunte's paper is evaluating market power based on welfare theory. Social welfare – or alternatively supply-chain performance – depends on two elements: (1) efficiency and (2) equity. Efficiency is concerned with the creation of value-added; equity is concerned with the division of value-added over the respective stakeholders. Efficiency and equity are not necessarily compatible. Efficient solutions may be very 'inequitable'. Maximizing value-added is not necessarily beneficial for all stakeholders concerned! First he addresses the question whether price changes at the farm level are fully and instantaneously transmitted into changes at the consumer level. He observes a general pattern of price asymmetry to the disadvantage of farmers and consumers. In general, price symmetry and price levelling are as prevalent as price asymmetry is. Secondly, he questions whether there have been changes in the price risk distribution in post-war agri-food supply chains. A shift in price risk from farmers to marketing organizations in the Dutch ware-potato supply chain has been observed. He finds limited evidence for the abuse of market power in European food processing and retail trade.

In the paper of Aramyan et al. performance indicators in agri-food chains are reviewed. The performance indicators range from highly qualitative indicators like customer satisfaction to quantitative indicators such as return on investments. Lacking consensus on what determines the performance of supply chains complicates the selection of one measurement system. Their literature review shows that many attempts have been made to develop a measurement system; however none have been successfully incorporated in practice. They suggest a performance measurement framework based on efficiency, flexibility, responsiveness and food quality.

The papers and discussion focused on one of the fundamental issues of chain performance: how are the objectives of chain and individual actor related? It is obvious that each stakeholder has his own objectives and his own performance measures. But it was also recognized that the performance of the chain as a whole is more than the sum of the performances of each individual chain member. That means that just one indicator will be insufficient for measuring chain performance, and the choice for performance indicators depends on the scope of performance measurement. Human capital, in particular the position of employees, is mentioned as a neglected item. Education, wages, productivity and job satisfaction could be performance measures in this case. Further research efforts are needed to connect theory and practice, but also to connect managerial opportunities with academic efforts to measure and improve chain performance. The issue of 'Who is the chain leader?' was also discussed. It is argued that in the end the supply-chain performance indicator should be directly related to the value-added for the final consumer. Also the position of retailers, their concentration and their influence on consumer prices as well as the distribution of the added value among the chain partners are important topics for further research.

SHARING COSTS, BENEFITS AND RISK IN AGRI-FOOD CHAINS

Integrated supply chains and networks, as a distinction from loosely related up- or downstream firms, offer opportunities for creating additional added value. High customer satisfaction and confidence in the purchase can be achieved by labelling or, at a higher level, branding. Labelling requires supply chains with a shared strategic focus and more chain coordination. Cooperation in chains enables traceability systems, which can reduce cost in the event of food safety problems and can improve verification of quality and credence values of food. Furthermore, the competitive positions of chains can be improved by using the competitive position of individual firms in the chain. In this section three papers based on empirical research are presented.

The growth of Minimum Quality Standards in the sector of fresh agricultural produce is a recent occurrence closely related to the food and food-safety crises of recent years. While the public authorities were creating new control and health-monitoring procedures, tightening regulatory production standards and enhancing regulations related to official marks of quality, some retailers were adopting new segmentation strategies for demand of a chain brand. Retailers are seeking to reassure consumers by creating their own quality labels and by communicating about the additional guarantees afforded to consumers by these quality labels. Giraud-Héraud and Soler discuss the impact on the performance of retailers and producers. The benefit of product differentiation through the implementation of a chain brand is higher for the retailer when the minimum quality level is low. The more the public authorities raise the minimum quality level the more difficult it becomes for retailers to implement a differentiation strategy. High levels of minimum quality standards make it impossible to establish a chain brand or private label. An increasing level of professionalism in the chain enables high standards, which decreases the possibilities of differentiation strategies.

Hobbs explores the economic functions of traceability, examining the extent to which traceability can bolster liability incentives for firms to practice due diligence. The extent to which consumers value traceability per se, versus verifiable quality assurances delivered through traceability, is evaluated empirically using survey and experimental auction data. The empirical analysis shows that consumers were willing to pay non-trivial amounts for a traceability assurance. For consumers, traceability has the highest value when bundled with additional quality assurances.

Ingenbleek uses the evolving logic in marketing to examine the problem of sharing financial rewards in agricultural supply chains. Building on resource-advantage theory it is suggested that the potential reward that firms may derive from participating in a supply chain depends on the competitive position of the chain as a whole and on the competitive position of the individual firm within the chain. To understand what its contribution to the chain is worth, the firm should be able to quantify relative customer value. Inappropriate assessments lead to a disability of the firm to take financial rewards in exchange for its contribution to the chain. To ensure that chain members remain motivated to invest in the chain and to provide them with sufficient financial resources to do so, it is in the common interest of all chain members that each of them is rewarded for its contribution to the competitive position of the chain.

The discussion focused on aspects of the retailers' strategy to create Premium Private Labels (PPL). Two main motivations for creating PPL are:

- Creating market power. Retailers experience severe price competition. Creating PPL enables higher profits, and also a higher market share can be one of the benefits. Furthermore, a wide range of PPL products enables economies of scale in advertising.
- Securing food safety. The General Food Law is a driver for food quality and safety standards. Some retailers apply additional requirements, thus creating a PPL. By coordinating the chain, retailers gain more influence on the production process and reduces the risk of insufficient food quality. Higher customer loyalty should outweigh higher chain costs due to food safety guarantees.

Despite clear conceptual approaches research on empirical verification needs to be strengthened.

Another issue discussed is the position of the supply base. A share in the higher profits, as showed by Giraud-Héraud and Soler, is not obvious. Retailers can buy their product from local producers but also from producers in low-cost countries. The value-added of producers participating in a PPL should be recognizable and therefore their produce should be differentiated from the bulk. A more traditional approach is creating countervailing power, e.g., by establishing a producers' cooperative. This cooperative can differentiate their produce and aim at different markets and at different countries.

MODELLING AGRI-FOOD CHAINS

The advantages of cooperation in agri-food supply chains and establishing consumer values by adding tangible and intangible assets to products can be argued from theoretical viewpoints. Modelling these advantages and assessing risks and robustness of chain cooperation provide information for decision making on chain cooperation. The papers in this session provide insight into the advantage of robustness modelling and the potential impact on chain profit by quality improvements and promotion activities.

Kleijnen[1] presented an approach of designing 'robust' as opposed to 'optimal' supply chains. The reason for such a robust approach lies in the stochastic nature of production and economic environments. Inherent risks and uncertainties prevailing in business environments may lead to sub-optimality of solutions in situations where, for example, key modelled parameters of the 'optimal' design differ from their real-life counterparts. As opposite, it is argued that robust solutions provide acceptable performance even when the number of possible scenarios is large. The variety of scenarios in this 'robust' framework is taken care of by assigning probability distributions to each scenario. The concept of robust supply chains is based on the ideas of Genichi Taguchi, a Japanese engineer who applied this approach to design Toyota cars. Kleijnen combined the approach of Taguchi with stochastic Monte Carlo simulation to incorporate the uncertainty into an analysis that provides robust solutions.

Schepers and Van Kooten present a quantitative model for the analysis of the optimal timing of selling commodities in agri-food retail chains. The key proposition of this paper is that larger gains for stakeholders in the chain can be achieved when fresh produce is sold at a riper stage. Inevitable high product losses and higher marketing costs will be outweighed by increased consumers' purchases of riper products. The model is dynamic and is based on three knowledge domains: consumer science, quality management and chain science. Applying the model shows how such actions as promotions and positioning of products influence the profits of stakeholders in the mango retail chain as well as the effects on cost-sharing agreements.

The discussion concerning the design of robust chains aimed at the question: "What are the specific robust issues facing developing optimal design of an agri-food supply chain?". It was agreed that these issues are largely determined by inherent characteristics pertaining to agricultural production and food distribution. The most important of these characteristics include the biological character of agricultural production, its close relation to nature and, hence, dependence on weather and other uncontrollable (and stochastic) natural forces, the perishability of products, and environmental concerns. A recent increase in public and scientific interest on 'tracking and tracing' issues in the supply chain, animal welfare and food safety concerns suggests these as important points to be taken into account when developing robust agri-food chains. Further research should therefore focus on incorporating the above issues in robust design of agri-food chains. The foreseeable challenge for those involved in this field of research may come from the difficulty of measuring important factors, such as food safety concerns, as well as notorious difficulties in getting (quantitative) data.

The discussion on the model presented by Schepers and Van Kooten centred on the possibilities for extending the model to deal with a lot of different agents, including producers, wholesalers, retailers and other stakeholders that are actually present in the chain. It was also stressed that any model should take into account the bargaining powers of the agents in the chain. The argument is that if a producer of, e.g., exotic fruits wishes to supply its products at a different ripeness stage than required by the retailer, the latter might switch to other suppliers whose produce conform to required specifications. Given that retail chains currently possess large bargaining power, efforts should be exercised to convince retailers that changes in the selling pattern proposed in the model may also benefit them.

THE VALUE OF INFORMATION IN AGRI-FOOD CHAINS

According to the US Department of Commerce, information technology has been one of the key drivers for economic growth in the USA: information technology was attributable for one third of real economic growth. In the 20th century mankind made a transition from an economy based on natural resources to one based on design and organization (Contractor and Lorange 2002). Due to this transition, the costs of carrying out exchanges between firms, whether in a market place or in a vertically integrated firm, increased. Growing diversity of knowledge sources, accelerating technical change, growing importance of outsourcing and hence a growing importance of vertical coordination are examples of the importance information and alliances play in a chain. In this section three papers are presented on management information infrastructure, on recalls in the dairy sector and on knowledge management in alliances.

Schiefer addresses the establishment and management of information infrastructures in chains. Tracking, tracing and quality-assurance systems in agriculture and the food sector are a prerequisite to support the guarantee of food safety and the focus on consumers' quality needs. The paper discusses the need for new information layers that utilize enterprise information but focused on the communication between chains for quality assurance towards the consumer and for improvements in risk management.

The General Food Law, which was implemented on 1 January 2005, emphasizes 'adequate traceability' and 'recalls whenever necessary'. The paper of Meuwissen et al. quantifies the costs of recall of consumption milk, and aspects for supporting due diligence are identified and prioritized. In counterattacking liability claims due diligence is central. Even when a company shows to be the cause of contamination, due diligence can demonstrate non-accountability and, hence, eligibility for insurance payments. Total recall costs are simulated using a Monte Carlo simulation model. Important parameters in the model are batch size and the point along the supply chain where contaminated products are identified. Analyses show that farmers are currently at a disadvantage in demonstrating due diligence compared to other participants of the chain.

Sporleder defines and analyses the concept of a strategic alliance as one specialized collaborative agreement among vertically allied firms in the supply chain. Intellectual property may serve as a base for *maximizing value-added* within a strategic alliance. Knowledge management provides novel insight into the foundations of a strategic alliance. The potential of a strategic alliance creating a real option for managers is examined along with the characteristics of networks that are organized around constant learning. The role of management is critical when evaluating strategic alliances. They need to be aware of what types of resources, tangible and intangible, are dedicated to the strategic alliance. Performance evaluation of alliances based on a certain-to-fuzzy continuum of inputs and outputs is suggested.

Concerning information management the discussion focused on the question what type of information is needed. In an empirical research for processed vegetables the end user, in this case the consumer, defines the needed information. In cases of liability it is important to know who is blamed according to the end user. The question is not only who is guilty but who has to repair the insufficient food quality. The performance of multilevel networks and chains can be measured by a balanced scorecard. Strategies concerning the financial, production, information and trust performance should be evaluated.

SUPPLY-CHAIN ORGANIZATION AND CHAIN PERFORMANCE

Markets are becoming increasingly global in scope, technologies are changing rapidly and the life cycles of products are ever-shortening. These changes in the external environment are incentives for cooperative activities between firms. Porter and Fuller (1986) mention the following benefits from cooperation:

- To achieve economies of scale and learning together with a partner.
- To get access to the benefits of other firms' assets: technology, market access, means of production, special skills.
- To reduce risk by sharing investments, including R&D.
- To speed up in reaching the market.

The papers in this section provide an overview of theoretical approaches of supply-chain organization and cooperation as well as empirical findings.

Wysocki et al. present the PWH (Peterson, Wysocki and Harsh) model, which focuses on vertical-coordination strategies of firms in agri-food chains. The model assesses the main decision points for strategy-making processes by firms with respect to vertical coordination. They analyse the chain on three phenomena: the firm's strategy to participate in the supply chain, the chain governance structure, and the application of industrial organization and institutional economic theory. In their overview they discuss the advantages of the various forms that agri-food chains may take: channel-master, chain-web and chain-organism model. They conclude that understanding agri-food chains requires a multidisciplinary approach, such as the PWH, learning supply-chain governance structure model.

The importance of chain transparency (e.g. market information symmetry) for developing corporate social responsibility (CSR) is emphasized by De Vlieger. The relationships between the social responsibilities of firms and chain organization are founded on the 'credence' characteristics of social responsibility. These credence characteristics make market information necessary to ensure market information is symmetric at every stage of the production chain. The incorporation of transparency on CSR into a chain is possible in a product-oriented or in a company-oriented way. De Vlieger supports his preference for the latter strategy with a company-oriented model in which the consequences of CSR for a chain are identified. The dynamic model addresses the opportunities for developing CSR by integrating it with quality-assurance systems, which are based on the same rules ('requirements').

Bijman presents a model for studying governance structure choice, applied to changes in governance structures in the Dutch fresh-produce industry. This industry has gone through substantial restructuring in recent years. Asset specificity has increased, mainly due to the introduction of brands and the establishment of specialized packaging stations. Measurement problems have increased due to specific quality attributes, product innovation and quality guarantees. As a solution, producers have vertically integrated by taking over or setting up downstream wholesale companies. Coordination problems have increased due to the shift from pooled to sequential interdependence. As a solution, coordination mechanisms have shifted from standardization to direct supervision. The latter has been materialized by giving the management of marketing cooperatives more authority

Relationships between key partnership characteristics and performance are discussed by Duffy and Fearne. In an empirical study they investigate how the development of more collaborative relationships between UK retailers and fresh-produce suppliers affect the financial performance of suppliers in such relationships. The results provide support for the theory that partnerships can help a firm to improve its performance. The results also show that commitment and trust and relational norms have the greatest predictive ability in the multiple-regression analysis, followed by functional conflict resolution and involvement in decisions and planning.

The discussion concentrated on concerns about variables that should be included in the different models. Although it was agreed that the PWH model reflects the important factors for adjusting or changing the strategy of a company, some variables seem to be interrelated. An example is the relationship between customer satisfaction and profit. In general, when the customer is satisfied the profits are higher. The CSR model shows a rather complex picture at first instance. Some variables have names that are difficult to understand. Could the content of the variable 'Chemistry' be the same as social capital? It seems that they have the same content because they are dealing with past experiences of a company in a chain; for example, the way of trading, communication and so on. The influence of norms/ standards on the type of governance structure has been questioned. It should be noticed that norms (values) vary across different nations. This gives an extra notion to collaboration in international food supply chains. In designing governance structure a trade-off has to be made between norms, authority and price mechanisms. The discussion also explored the issue of transparency. Will these models achieve more transparency in international food supply chains? Are members willing to give more insight into their business processes and are they willing to give (sensitive) information for more transparency in chains? And how will the quality of the transparency variables be guaranteed? The papers and discussion made it clear that the theories give the concepts, but that each chain has to deal with these issues separately as a part of the chain cooperation. Next, the practical value of the models was discussed. In the PWH model a path is made to achieve the appropriate strategy for a company or a chain. Is it possible to use this model for optimizing the strategy and could this strategy be placed on the continuum for the different forms of learning supply chains? Does the cooperative model only give a situation of steady state or is the model also able to take care of dynamism in the business environment or, in other words, can the model determine the best structure for the chain of tomorrow?

CONCLUSIONS

The emerging field of quantifying the agri-food supply chain is demonstrated by the chapters in this volume. First, a wide variety of approaches from different economic disciplines are useful in analysing the supply chain. The approaches tackle questions to get optimal, efficient or robust solutions of chain performance, structure, organization or chain strategy. Concepts, cases and empirical findings of these approaches in agri-food supply chains have been demonstrated. Second, understanding the complex systems of agri-food chains requires more investments in retrieving empirical data for testing propositions and developing appropriate models. The experiences from past research can be and should be further explored. This will result in opportunities and support managerial decisions. Third, the contributions show that on an international level large interest exists in the issue of supply chains and in the importance of the agri-food chain. The identified research gaps and discussion points are shared among an international forum of researchers. International cooperation among researchers will enhance progress in this research field. The workshop was highly valued in this respect.

ACKNOWLEDGEMENT

The authors would like to thank the PhD students Michael Gengenbach, Derk-Jan Haverkamp, Tatiana Novoselova, Wijnand van Plaggenhoef, Ilya Surkov and Bouda Vosough-Ahmadi for summarizing the discussions.

NOTES

[1] This presentation is not included in this volume.

REFERENCES

Contractor, F.J. and Lorange, P., 2002. The growth of alliances in the knowledge based economy. *In:* Contractor, F.J. and Lorange, P. eds. *Cooperative strategies and alliances*. Pergamon, Amsterdam, 3-22.

Kaplan, R.S. and Norton, D.P., 1992. The balanced scorecard: measures that drive performance. *Harvard Business Review,* 70, 71-79.

Porter, M.E. and Fuller, M.B., 1986. Coalitions and global strategy. *In:* Porter, M.E. ed. *Competition in global industries*. Harvard Business School Press, Boston, 315-343.

MEASURING PERFORMANCE IN AGRI-FOOD CHAINS

CHAPTER 2

PERFORMANCE MEASUREMENT IN AGRI-FOOD SUPPLY-CHAIN NETWORKS

An overview

JACK G.A.J. VAN DER VORST

Logistics and Operations Research Group, Wageningen University, Hollandseweg 1, 6706 KN Wageningen, The Netherlands

Abstract. Many researchers and practitioners are working on the enhancement of supply-chain collaboration in order to improve performance of the individual supply-chain members and supply-chain performance as a whole. Performance measurement fulfils a crucial role in the development of supply chains as it can direct the design and management of the chain towards the required performance. It is the key instrument to discuss and evaluate the effectiveness of (potential) chain partnerships. This paper presents a framework for the development of innovative food supply-chain networks and discusses the implications for performance measurement systems. Current bottlenecks and research opportunities are presented.
Keywords: performance indicators; network optimization; (bottlenecks and) research opportunities

INTRODUCTION

The importance of performance measurement has long been recognized. Manufacturing and management consultant Oliver Wright almost 30 years ago offered the oft-repeated saying, "You get what you inspect, not what you expect" (Melnyk et al. 2004). Metrics are therefore needed to evaluate how work is done and to direct the activities, since what we measure indicates how we intend to deliver value to our customers. Incorrect performance measurement systems (PMS) can create disincentives and unwanted behaviour.

The number of publications on performance measurement has increased significantly in the last decade (e.g. Beamon 1999; Lohman et al. 2004; Gunasekaran et al. 2004). This is mainly because of a number of fundamental changes in the business environment, especially in agri-food chains. Consumers in Western-European markets have become more demanding and place new demands on attributes of food such as quality (guarantees), integrity, safety, diversity and associated information (services). Demand and supply are no longer restricted to

C.J.M. Ondersteijn et al. (eds.), Quantifying the agri-food supply chain, 15-26.

nations or regions but have become international processes. We see an increasing concentration in agribusiness sectors, an enormous increase in cross-border flows of livestock and food products and the creation of international forms of cooperation. The food industry is becoming an interconnected system with a large variety of complex relationships, reflected in the market place by the formation of (virtual) Food Supply Chain Networks (FSCNs) via alliances, horizontal and vertical co-operation, forward and backward integration in the supply chain and continuous innovation (Beulens et al. 2004). The latter encompass the development and implementation of enhanced quality, logistics and information systems that enable more efficient execution of processes and more frequent exchange of huge amounts of information for coordination purposes (Van der Vorst et al. 2005). All these developments initiate a reorientation of companies in Dutch agriculture and food industry on their roles, activities and strategies. As a consequence also the PMS need adjustments as traditional measurement approaches may limit the possibilities to optimize the FSC(N) as a whole.

This paper presents an overview of performance measurement in agri-food supply-chain networks. We will first go deeper into the concept of FSCN. Next, we will discuss a framework for chain/network development that will be used to derive requirements for PMS. We will discuss bottlenecks of performance measurement in FSCN and conclude with an overview of research opportunities in this area.

FOOD-SUPPLY-CHAIN NETWORKS

Supply-chain management (SCM) is the integrated planning, coordination and control of all business processes and activities in the supply chain to deliver superior consumer value at least cost to the supply chain as a whole while satisfying the variable requirements of other stakeholders in the supply chain (e.g. government and NGOs) (Van der Vorst 2000). In this definition a *supply chain* is a series of (physical and decision-making) activities connected by material and information flows and associated flows of money and property rights that cross organizational boundaries. The supply chain not only includes the manufacturer and its suppliers, but also (depending on the logistics flows) transporters, warehouses, retailers, service organizations and consumers themselves. In the definition of SCM a *business process* refers to a structured, measured set of activities designed to produce a specified output for a particular customer or market (Davenport 1993). Next to the logistical processes in the supply chain (such as operations and distribution) we distinguish business processes such as new-product development, marketing, finance, and customer relationship management (Chopra and Meindl 2001). Finally, *value* is first of all the amount consumers are willing to pay for what a company provides and it is measured by total revenue. The concept 'value-added activity' originates from Porter's 'value chain' framework and characterizes the value created by an activity in relation to the cost of executing it (Porter 1985). Currently the value concept is more expanded. We now talk about values associated with the so-called 'Triple P': People, Planet and Profit. So, next to financial performance also social and environmental performance are incorporated. These latter two lead to (qualitative) attributes that are generally spoken associated with the product itself (biologically produced), the companies producing it (social policy) and the raw materials (GMO?) and resources (child labour?) used.

Figure 1 depicts a generic supply chain at the organization level within the context of a complete supply-chain network. Each firm is positioned in a network layer and belongs to at least one supply chain: i.e. it usually has multiple (varying) suppliers and customers at the same time and over time. Other actors in the network influence the performance of the chain. As Håkånsson and Snehota (1995) state: "what happens between two companies does not solely depend on the two parties involved, but on what is going on in a number of other relationships". Therefore, the analysis of a supply chain should preferably take place or be evaluated within the context of the complex network of food chains, in other words a Food Supply Chain Network (FSCN). Lazzarini refers to a 'netchain' and defines it as "a directed network of actors who cooperate to bring a product to customers" (Lazzarini et al. 2001).

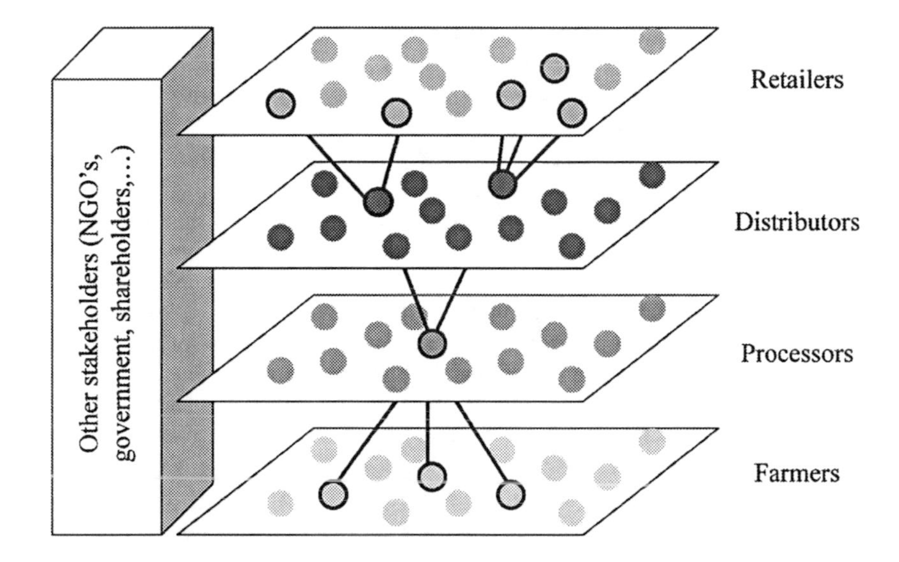

Figure 1. *Schematic diagram of a supply chain from the perspective of the processor (bold flows) within the total FSCN (based on Lazzarini et al. 2001)*

In an FSCN different companies collaborate strategically in one or more areas while preserving their own identity and autonomy. As stated, in an FSCN more than one supply chain and more than one business process can be identified, both parallel and sequential in time. As a result, organizations may play different roles in different chain settings and therefore collaborate with differing chain partners, who may be their competitors in other chain settings. In brief, chain actors may be involved in different supply chains in different FSCNs, participate in a variety of business processes that change over time and in which dynamically changing vertical and horizontal partnerships are required. This puts stringent requirements on the PMS as we will discuss later.

FRAMEWORK FOR CHAIN/NETWORK DEVELOPMENT

When researchers and/or managers discuss the potentials of chain and network development, there is a need for a 'language', a framework, that will allow us to describe supply chains, its participants, processes, products, resources and management, relationships between these and (types of) attributes of these in order to allow us to understand each other unambiguously (to a large extent). This section presents such a framework (Van der Vorst et al. 2005).

In an FSCN a number of typical characteristics can be identified. In line with the thoughts of Lambert and Cooper (2000) we distinguish the following four elements that can be used to describe, analyse and/or develop a specific (supply chain within the) FSCN (see Figure 2):

1. *The Network Structure* demarcates the boundaries of the supply-chain network and describes the main participants or actors of the network, accepted and/or certified roles performed by them and all the configuration and institutional arrangements that constitute the network. The key is to sort out which members are critical to the success of the company and the supply chain – in line with the supply-chain objectives – and, thus, should be allocated managerial attention and resources.
2. *Chain Business Processes* are structured, measured sets of business activities designed to produce a specified output (consisting of types of physical products, services and information) for a particular customer or market. As stated before, next to the logistical processes in the supply chain (such as operations and distribution) we distinguish business processes such as new-product development, marketing, finance, and customer relationship management.
3. *Network and Chain Management* typifies the coordination and management structures in the network that facilitate the instantiation and execution of processes by actors in the network, making use of the chain resources with the objective to realize the performance objectives formulated by the FSCN. Lambert and Cooper (2000) distinguish two groups of management components (see Table 1). Especially the managerial and behavioural components are well-known obstacles to SCM as they might hinder the development of trust, commitment and openness between supply chain members.
4. *Chain Resources* are used to produce the product and deliver it to the customer (so-called transforming resources). These enablers include people, machines and ICT (information, information systems and information infrastructures).

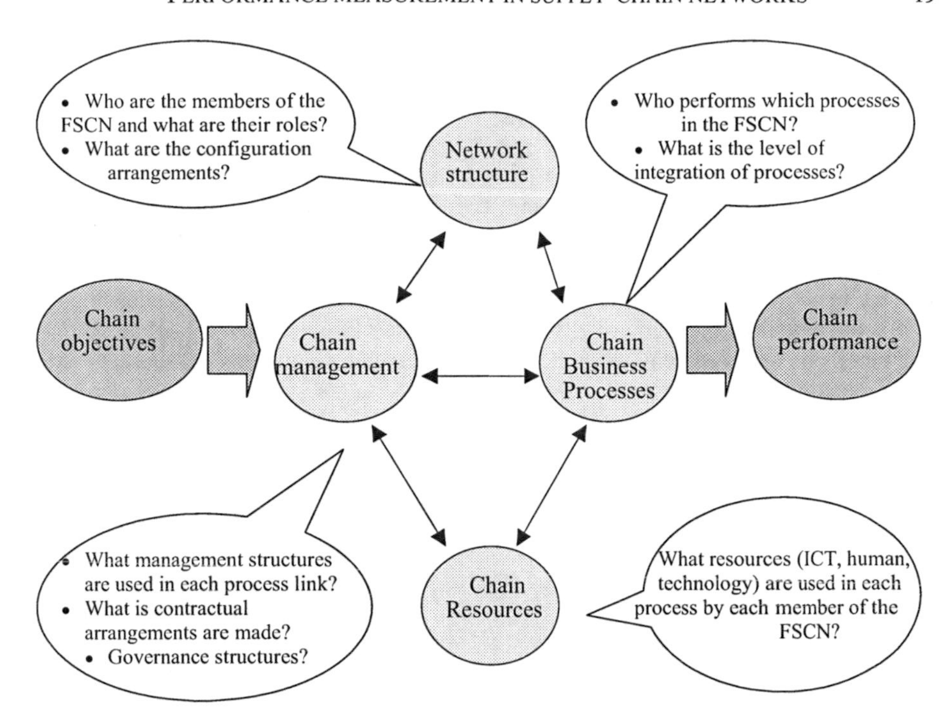

Figure 2. *Framework for chain/network development (adapted from Lambert and Cooper 2000)*

Table 1. *Two groups of management components that have to be aligned in the supply chain*

Physical and technical components	Managerial and behavioural components
• planning and control methods (e.g. push or pull control) • work flow/activity structure (indicates how the firm performs its tasks and activities) • organization structure (indicates who performs the tasks and activities, e.g. cross-functional teams) • communication and information flow facility structure (e.g. information transparency) • product flow facility structure (e.g. location of inventories, decoupling points)	• management methods (i.e. the corporate philosophy and management techniques) • corporate culture and attitude • risk and reward structure • power and leadership structure

In brief, within an FSCN we identify one or more Chain Business Processes with well-identified products that are produced and delivered to the customer of that Chain Business Process (e.g. chain processes that produce and deliver boxes with yoghurt or cheese slices to retail outlets). These production and delivery processes require the execution of *business activities* by one of the actors participating in the network (such as transportation, storing, order picking). There are precedence relations between business activities that are or may be determined by goods, resources, information, financial and control flows. So we may regard an FSCN as a directed network of business processes and activities with precedence relationships.

Each element of the framework is directly related to the *objectives of the FSCN*. One can focus on three generic value propositions, which can be found separately or in combination:

1. Network differentiation and market segmentation where the target is to differentiate as a chain to meet the specific demands of customers (e.g. assortment, product quality, etc.).
2. Integrated quality; the target here is to meet the increasing demand of consumers, governments, NGOs and business partners for safe and environmentally friendly produced products.
3. Network optimization; the target here is cost reduction through a streamlined and efficient chain/network with rational information supply.

Whether these objectives are realized in practice can be measured via output performance of the supply chain (network). Supply-chain performance is defined as the degree to which a supply chain fulfils end-user and stakeholder requirements concerning the relevant performance indicators at any point in time. *Performance indicators* (or performance metrics) are operationalized process characteristics, which compare the performance of a system with a norm or target value. Or, as Christopher (1998) states, "they refer to a relatively small number of critical dimensions which contribute more than proportionally to the success or failure in the marketplace". It depends on the objectives of the supply chain as to which specific *key performance indicators* (KPIs) are appropriate and used.

The groundwork for successful SCM is established by an explicit definition of supply-chain objectives and related KPIs and, successively, by deciding on the four key elements of the FSCN. The optimal design will differ for each supply chain depending on the competitive strategy and the market, product and production characteristics.

IMPLICATIONS FOR PERFORMANCE MEASUREMENT

Performance measurement aims to support the setting of objectives, evaluating performance, and determining future courses of action on a strategic, tactical and operational level. To meet objectives, the output of processes must be measured and compared with a set of standards. In order to be controlled, the process parameter values need to be kept within a set limit and remain relatively constant. This will allow comparison of planned and actual parameter values and taking certain reactive measures in order to improve the performance or re-align the monitored value to the defined value (Gunasekaran et al. 2004).

Since a supply chain is by its definition a collection of multiple actors with each their own specific objectives (and values and norms) a lot of effort has to be put in the development of a shared language, shared objectives, shared KPIs, etc. A well-defined set of chain performance indicators will help establish benchmarks and assess changes over time; but only when all stages in the supply chain aim to realize the same jointly defined objectives. As we have described the agri-food industry is becoming an interconnected system with an even larger variety of complex relationships reflected in the market place by the dynamic formation of chain partnerships. Chain actors may be involved in different supply chains and participate in a variety of business processes that change over time and in which dynamically changing vertical and horizontal partnerships are required. This places very specific (and dynamic) requirements on the PMSs of such companies, requiring high flexibility and possibilities for making integral analyses.

The supply-chain performance is an overall performance measure that depends on the performances of the individual chain stages and the respective processes that are executed in those stages. Processes can necessarily be identified at different levels of abstraction. That means that each process can be broken down into a directed and connected network of (sub-)processes/activities. As a consequence one makes a conscious choice of the abstraction level needed in the business context. As an example, (Van der Vorst 2000) presents a framework of logistics performance indicators that is divided into three hierarchical decision levels, namely the supply-chain performance, the performance of an individual organization and the performance of an individual business process (Table 2). All indicators are composites of, and dependent on, lower-level measures. For example, the supply-chain lead time and product quality are dependent on the throughput times of business processes in all chain/network stages.

While traditional PMSs are based on costing and accounting systems, measuring performance in supply-chain networks requires a more balanced set of financial and non-financial measures at various points along the supply chain (Lohman et al. 2004). A relevant development is the balanced-scorecard approach, which includes these additional performance sets (Kaplan and Norton 1992). But more development is needed. Next to traditional performance indicators such as costs, throughput time or technical quality of products also other indicators in line with the 'Triple P' philosophy (People, Planet, Profit) have to play a role in this process and should therefore be developed; examples are 'guaranteed product integrity', 'environmental chain profile' or 'profile of animal-friendliness' and 'guaranteed quality, hygiene and safety' in meat-producing chains. As stated, the chain/network objectives should play a directive role in this selection and definition process.

Table 2. *Example of Logistic KPIs for food-chain networks on three hierarchical levels (Van der Vorst, 2000)*

Level	Performance indicator	Explanation
Supply-chain network	Product availability on shelf	Presence of a large assortment and no stock-outs
	Product quality	Remaining product shelf life
	Responsiveness	Order cycle time of the SC
	Delivery reliability	Meeting guaranteed delivery times
	Total SC cost	Sum of all organizations' costs in the supply chain
Organization	Inventory level	Number of products in store
	Throughput time	Time needed to perform chain of business processes
	Responsiveness	Flexibility of the organization: lead time
	Delivery reliability	% Orders delivered on time and in right quantity
	Total organization's cost	Sum of all process costs in the specific organization
Process	Responsiveness	Flexibility of the process
	Throughput time	Time needed to perform the process
	Process yield	Outcome of the process
	Process cost	Cost made when executing the process

There is a need to define and measure performance for each instantiation of the supply-chain network as a whole and to the level of the participating organizations and executed processes. Therefore, in line with the framework presented in the previous section, the set up of the PMS requires the identification of:

- A balanced number of performance metrics at multiple aggregation levels departing with the network objectives to capture the essence of the chain and organizational performance (Gunasekaran et al. 2004). This means taking into account:
  - indicators for the *chain network structure* to benchmark the objectives of each member. Is the chain/network to be evaluated on environmental issues or just financial performance? Examples of indicators are the contribution of each member in the total added value, ROI, etc.;
  - indicators for the output of the relevant *chain business processes* and *chain management structure* to assess the efficiency and effectiveness of the planning and control activities (e.g. logistics metrics such as lead time, responsiveness, inventory levels, delivery reliability, product quality, etc.);

  - indicators at process level related to *chain resources* utilization (e.g. process yield, utilization degree), well-being (humans) and perseverance (fit for the future).
- Dynamic metrics that recognize and respond to changes in customer requirements, operating inputs, resources and performance over time, and identify and anticipate to potential problems proactively;
- Metrics that fulfil the well-known evaluation criteria, such as validity (accurateness), robustness (similar interpretation by all users, repeatable, comparable across time and place), usefulness (understandable, benchmarkable and providing a guide for action), economy (cost–benefit evaluation of collecting and analysing the data), inclusiveness (measurement of all pertinent aspects), verifiable (based on an agreed upon set of data and a well-understood and well-documented process for converting these data into the measure) and consistency (measures consistent with organizational goals) (a.o. Melnyk et al. 2004; Beamon 1999; Caplice and Sheffi 1994).

BOTTLENECKS OF, AND DEVELOPMENTS IN PERFORMANCE MEASUREMENT

There is a need to define and measure performance for the supply chain as a whole and to be able to drill down to different measures and different levels of detail, in order to understand the causes of significant deviations of actual performance from planned performance (Lohman et al. 2004). However, many companies seem to be facing serious difficulties in developing and implementing such supply-chain-wide PMSs that capture various dimensions of performance at various levels in a consistent way. This was confirmed in a recent international study on traceability systems in agri-food supply chains where we found a general lack of chain cooperation and transparency (Van der Vorst 2004). Let's take a closer look at some of the bottlenecks:

- There is often a history of decentralized reporting with a focus on local operational use within factories, distribution centres, etc. This has led to an uncontrolled growth of reports with many inconsistencies, which have to do with definitions of performance metrics, sources of data for obtaining measures, and ways of presenting reports (Lohman et al. 2004). These local metrics, data gathering and reporting structures hinder an integrated analysis.
- There is a lack of standard definitions of KPIs and measurement methods. When companies start working in a supply-chain concept they often speak a different language; their objectives and definitions of KPIs are not harmonized. This is also due to the divergence in value propositions in the FSCN (note that practically each supply-chain member does business with multiple suppliers and customers).
- There is divergence in development stages of organizations. Some organizations lack internal integration and still have functional silos that do not cooperate. Furthermore, some have very sophisticated electronic information infrastructures whilst others have huge paper archives (divergence in ICT development phases).

Companies use many information systems that are linked in some way. The dispersed IT infrastructure produces a number of issues (Lohman et al. 2004):

- it adds to the lack of data integrity between the reports. Since considerable overlap exists between the systems, certain data can be extracted from multiple sources and this often leads to inconsistency;
- the infrastructure does not provide visibility over the supply chain, owing to the absence of connectivity;
- certain systems are not designed for reporting uses or cannot provide data at reasonable cost at all.

More and more multinationals practice benchmarking of production and sales units and have developed scorecards. Lately, more work is done on the standardization of performance metrics and PMS. For example, the Supply-Chain Council developed the Supply Chain Operations Reference (SCOR) model; a process reference model developed as the cross-industry standard diagnostic tool for SCM based on the basic processes 'Plan, Source, Make, and Deliver'. KPIs within SCOR focus on supply-chain delivery reliability, responsiveness, flexibility, costs and asset management. However, often a scorecard tailored to the needs of the specific company, chain or network is needed. Lohman et al. (2004) emphasize that the development of a PMS should be considered a co-ordination effort rather than a design effort. They suggest the development of a metrics dictionary using the metrics definition template presented in Table 3 as the main element in the

Table 3. *Performance metrics definition template (Lohman et al. 2004)*

Metric attribute	Explanation
Name	Use exact names to avoid ambiguity
Objective	The relation of the metric with the organizational objectives must be clear
Scope	States the area of business or parts of the organization that are included
Target	Benchmarks must be determined in order to monitor progress
Equation	The exact calculation of the metric must be known
Units of measure	What is/are the unit(s) used
Frequency	The frequency of recording and reporting of the metric
Data source	The exact data sources involved in calculating a metric value
Owner	The responsible person for collecting data and reporting the metric
Drivers	Factors that influence the performance, i.e., organization units, events, etc.
Comments	Outstanding issues regarding the metric

development of a PMS. Preferably, this development departs with existing reports at various levels in the organization(s) to understand current metrics in detail, to identify shortcomings, and to include ongoing initiatives that affect PM (such as new information systems, etc.). The method will develop metrics in a consistent way and identify gaps in the current selection of metrics when confronted with the organizational objectives.

RESEARCH OPPORTUNITIES IN PERFORMANCE MEASUREMENT IN FSCN

Melnyk et al. (2004) point out that the topic of metrics as discussed by managers differs from the topic of measurement as typically discussed by academics. This is because academics are concerned with defining, adapting and validating measures that can be generalized to address specific research questions, whereas managers are generally more than willing to use a 'good enough' measure if it can provide useful information quickly. But when is the measure good enough?

Each FSC(N) requires its own PMS depending on the strategy and the FSC(N) characteristics. There is a need for the development of a balanced (dynamic) set of financial and non-financial FSCN performance indicators that reflect the inter-dependencies of different areas at the right aggregation level. There is a need for standard definitions of performance indicators to allow for integral analyses in dynamic configurations of FSCN. New KPIs are needed on different aggregation levels, because "PI's wear out as a result of their successful use as people adapt to the way they find themselves being measured and evaluated" (also known as the Hawthorn effect). Other research questions for the near future are:

- What metric set is suitable for the four levels in the framework for chain development?
- Should all metrics be mathematically derived or is there room for qualitative metrics?
- How should one make the trade-off between financial metrics and non-financial metrics?
- What environmental and social performance metrics can be developed that meet the requirements of the consumer?
- What is the optimal size of a metric set and how does one derive a predictive metric set?
- If one is to obtain proactive control, what predictive metric set (financial/non-financial) is suitable for what situation?
- How can we model in a generic way the dynamic configurations and performances of FSCN that concern more links and incorporate the requirements of all stakeholders in the FSCN? What quantitative method or technique is applicable? (see, e.g., Kleijnen and Smits 2003)
- How can we make these modelling methods comprehensible for managers so that the outcomes will be accepted?

REFERENCES

Beamon, B.M., 1999. Measuring supply chain performance. *International Journal of Operations and Production Management,* 19 (3/4), 275-292.

Beulens, A.J.M., Coppens, L.W.C.A. and Trienekens, J.H., 2004. *Traceability requirements in food supply chain networks*. Wageningen University, Wageningen. Working Paper.

Caplice, C. and Sheffi, Y., 1994. A review and evaluation of logistics metrics. *International Journal of Logistics Management,* 5 (2), 11-28.

Chopra, S. and Meindl, P., 2001. *Supply chain management: strategy, planning, and operation*. Prentice Hall, Upper Saddle River.

Christopher, M., 1998. *Logistics and supply chain management: strategies for reducing costs and improving services*. 2nd edn. Financial Times/Prentice Hall, London.

Davenport, T.H., 1993. *Process innovation: reengineering work through information technology*. Harvard Business School Press, Boston.

Gunasekaran, A., Patel, C. and McGaughey, R.E., 2004. A framework for supply chain performance measurement. *International Journal of Production Economics,* 87 (3), 333-347.

Håkånsson, H. and Snehota, I., 1995. *Developing relationships in business networks*. Routledge, London.

Kaplan, R.S. and Norton, D.P., 1992. The balanced scorecard: measures that drive performance. *Harvard Business Review,* 70, 71-79.

Kleijnen, J.P.C. and Smits, M.T., 2003. Performance metrics in supply chain management. *Journal of the Operational Research Society,* 54 (5), 507-514.

Lambert, D.M. and Cooper, M.C., 2000. Issues in supply chain management. *Industrial Marketing Management,* 29 (1), 65-83.

Lazzarini, S.G., Chaddad, F.R. and Cook, M.L., 2001. Integrating supply chain and network analyses: the study of netchains. *Journal on Chain and Network Science,* 1 (1), 7-22.

Lohman, C., Fortuin, L. and Wouters, M., 2004. Designing a performance measurement system: a case study. *European Journal of Operational Research,* 156 (2), 267-286.

Melnyk, S.A., Stewart, D.M. and Swink, M., 2004. Metrics and performance measurement in operations management: dealing with the metrics maze. *Journal of Operations Management,* 22 (3), 209-217.

Porter, M.E., 1985. *Competitive advantage: creating and sustaining superior performance*. Free Press, New York.

Van der Vorst, J.G.A.J., 2000. *Effective food supply chains: generating, modelling and evaluating supply chain scenarios*. Proefschrift Wageningen [http://www.library.wur.nl/wda/dissertations/dis2841.pdf]

Van der Vorst, J.G.A.J., 2004. Performance levels in food traceability and the impact on chain design: results of an international benchmark study. *In:* Bremmers, H.J., Omta, S.W.F., Trienekens, J.H., et al. eds. *Dynamics in chains and networks: proceedings of the sixth international conference on chain and network management in agribusiness and the food industry (Ede, 27-28 May 2004)*. Wageningen Academic Press, Wageningen, 175-183.

Van der Vorst, J.G.A.J., Beulens, A.J.M. and Van Beek, P., 2005. Innovations in logistics and ICT in food supply chain networks. *In:* Jongen, W.M.F. and Meulenberg, M.T.G. eds. *Innovation in agri-food systems: product quality and consumer acceptance*. Wageningen Academic Publishers, Wageningen, 245-292.

CHAPTER 3

ACCOUNTING STANDARDS FOR SUPPLY CHAINS

HARRY BREMMERS

Business Administration Group, Wageningen University, P.O. Box 8130, 6700 EW Wageningen, The Netherlands. E-mail: harry.bremmers@wur.nl

Abstract. In this paper principles for accounting in supply chains will be developed. The three principles we introduce are: the reciprocity in information access, asset investment and retrieval, and matching risks and returns. The principles provide guidance in managerial and stakeholder decision making, monitoring and control. The information that management and other stakeholders need is based on these standards.

The use of the supply-chain accounting principles increases transparency. Transparency is one of the main elements of what Fowler et al. (2004) call 'virtually embedded ties'. We argue that transparency should replace, or at least supplement, relational trust in supply channels. We argue also that technological innovation, such as electronic chain-wide reporting, is beneficial for transparency, decision making and control in/of supply chains, and will reduce the administrative costs of (supply-chain) accounting system at the same time.
Keywords: financial accounting; reporting guidelines; accounting standards; reciprocity

INTRODUCTION AND OUTLINE

In this paper we ask ourselves: (1) does the Dutch legal framework permit supply-chain accounting and (2) for what items (problem areas) is it necessary? and (3) what principles should be applied? Given the lack of research in this area and the presence of single-business accounting standards, the development of supply-chain accounting principles in this paper (1) is a normative rather than an objective enterprise and (2) uses the available standards that have already been developed in accounting theory and practice. In the section on accounting we state that the Anglo-American reporting framework allows for chain-wide reporting better than Roman-based accounting systems, which are existent in France and Germany. Although Roman influences also have permeated in the Dutch reporting structure, it has been influenced more by the Anglo-American framework. The Anglo-American standards are less rigid and more purpose-oriented than the fiscally-minded Roman structure is.

Chain-wide accounting is necessary to assess and account for social, environmental and economic performance parameters (also known as People–Planet–Profit or PPP performance). We oppose the traditional systems of accounting

C.J.M. Ondersteijn et al. (eds.), Quantifying the agri-food supply chain, 27-38.

and performance measurement and argue that it is necessary to develop accounting rules for chain-wide reporting and disclosure. In the next section we set forward three principles of supply-chain accounting. In the final section (on Information processing and innovation) we argue that the development of supply-chain accounting systems should be supplemented by legal and technological innovation, to increase transparency and efficiency in information processing as well as to reduce to the administrative burden of companies.

ACCOUNTING APPLIED SYSTEMS AND STRUCTURE

In the past, accounting, especially financial accounting, has mainly focused on the *ex-post* quality of information provision by the individual firm (see, for instance, Drury 1992). The business corporation, as a legal entity, has as characteristics: entity, accountability and independence. The entity concept refers to Ijiri's conception of the firm as having rights and obligations different from its financial stakeholders. The distance between the two is bridged by the provision of audited information by management. Management governs the property rights of the firm. Property rights can be *legal property* and *economic rights*.

Economic rights (and related obligations) refer to contracts and constructions to guarantee the exploitation of the economic benefits related to an asset. Supply-chain assets (collective structures for the processing of products) are often economic rather than legal of nature. Not all reporting systems support the disclosure of such assets. As a matter of fact, the way financial accounting systems deal with economic rights is quite different within the European Union. In The Netherlands, which has a reporting system that is strongly influenced by an Anglo-American viewpoint, economic rights held by an individual firm can be disclosed as assets on a balance sheet, provided that the 'assets' have distinct features that are similar to legal property. Under Dutch reporting rules, which hold the middle between an Anglo-American and a Roman system of disclosure, assets are recognized in the published accounts only if they can be traded, can be isolated for individual valuation, and whose value can be assessed in concordance with acceptable principles (see, for instance, Choi et al. 2002; Bremmers 1995). Financial leases, for instance, are regarded as assets under Dutch law. This system safeguards the main function of financial reporting: giving a true and fair view of the financial position of the firm, i.e. of its assets and liabilities, profits and losses as well as of its cash position. Because of this, the Dutch legal system would support the provision of information on the performance, the assets and liabilities of a supply chain if, and only if, this leads to a better insight into the financial position. On a French balance sheet only legal property is accounted for, while in Germany (because of the application of the 'Maßgebligkeitsprinzip') generally only those assets and liabilities that are disclosed are fiscally acceptable.

Accounting is divided into management accounting and financial accounting. Management accounting focuses at the internal, managerial decision-making process and control of the firm (Drury 1992). It contributes to responsible and accountable governance. Financial accounting on the other hand focuses at the provision of information to external stakeholder groups, of which the shareholders are the most important (see, for instance, Wubben and Bremmers 2003). Both accounting methodologies stimulate the creation of shareholder value. The relationship between shareholders and management is one between principals and agents. The agents (management) should, in a world with asymmetrically distributed information, respond to the need for transparency of different stakeholder groups (principals). Accounting, whether it is management accounting or financial accounting, creates the needed transparency and accountability. It does so by composing and sending messages, and therefore can be depicted as a *communication device*. Communication is defined as interaction through messages. In the words of Shannon and Weaver (1949, cited in Alter and Roche 1999) communication is "all procedures by which one mind may affect another". Since accounting is "the art of communicating financial and non-financial data", according to Shannon and Weaver three problem areas exist:

- technical: achieving efficient transmission and reception of data;
- semantic: increasing precision of message transfer;
- effectiveness: level of behavioural influence.

Improvement in these areas could reduce the level of information asymmetry. Information asymmetry exists, among other things, with respect to performance assessment: environmental performance (Planet), social performance (People) and financial performance (Profit). The Balanced Scorecard approach would depict the level of performance in the areas depending on the efficacy and efficiency of processes (channels), learning and growth activities and consumer satisfaction. These are also the main topics in supply-chain management. This means that supply-chain reporting should address the main performance areas, as well as the causal relationships between PPP performance and channel structure and processes.

Environmental performance disclosure is necessary for several reasons. Legal obligations (public policy), valuation of changes in product configuration, as well as necessary changes in perceived utility of environmental improvements all require some form of environmental performance measurement (Bremmers 2000; 2001). Since it can be expected that in practice environmental measures are taken more willingly if 'pollution prevention pays', or 'the polluter is made to pay', environmental performance indicators should inform about the consequences of alternate environmental strategies (Bremmers et al. 1996). This is not only true for a single firm but also holds for an entire supply chain, as is the case in life-cycle analysis (LCA). Supply-chain environmental efforts are not only accompanied by costs (Buzzelli 1991) but also reduce costs (Madsen and Ulhøi 2001; Tyteca et al. 2002) and possibly create first-mover advantages (e.g. Welford and Gouldson 1993; also: Carroll 1979; Wartick and Cochran 1985; Clarkson 1995; Waddock and Graves 1997; Husted 2003).

Social performance disclosure is relevant because of the fact that human and social assets are of more importance for the individual firm than are easy-to-replace tangible assets, like machines or buildings (Sporleder and Peterson 2003). Nowadays, social responsibility of companies is not limited to increasing wealth but stretches out towards accomplishing this in a socially and environmentally sustainable setting (Friedman 1970; Beamon 1999; Hart 1995). This is emphasized by the development of Corporate Social Responsibility (CSR) in the corporate world.

Financial performance is traditionally measured by the amount of profits that is made by individual firms. Since financial performance measurement has a long history both in practice and financial literature, this measure is scrutinized in the following paragraph.

Traditional accounting systems and performance measurement

Traditional accounting systems focus on static, isolated and *ex-post* profit measurement. Modern performance measurement instruments on the other hand, include a multitude of areas of performance, related to the different (primary) stakeholder groups. In our view, supply-chain accounting standards will have to include multiple performance measurements at different levels: operational, managerial, logistic and at a communicational level. The specifics depend on the supply-chain structure. A traditional supply-chain structure is centralized with respect to strategic decision making and control and focuses on a single-company profit increase. Power exertion is the coordination mechanism that is applied. On the other hand, a supply-chain governance system that focuses on relational equality (cooperative supply-chain governance) will try to meet different goals of stakeholder groups that are involved. At the far end of the continuum, an atomic supply chain will leave all coordination to the market.

The present disclosure of the *ex-post* profitability by individual firms, as is exercised under commonly accepted accounting standards, is based on neo-classical economic theory. Neo-classical economic propositions include the availability of homogeneous products, the absence of influence on prices by individual firms, and the existence of many competing suppliers. Under these propositions, performance measurement instruments are output-oriented and profit-related (Acs and Gerlowski 1996). The focus on single-firm profitability has major disadvantages (Bremmers 2001).

First of all, profit measurement is carried out retrospectively. The accounting concept of profit that is generally applied measures the equity at two moments in time, to assess profitability. The precision of this type of profit measurement can be questioned, since profitability can easily be influenced by the asset valuation system. For decision making by stakeholders, a Hicksian (dynamic) *economic* concept of profit would, therefore, be more appropriate.

Secondly, the accountant's profit measurement lacks the inclusion of risk as a significant part of business performance (compare Hardaker et al. 1997). Awareness of food-safety issues, due to recent crises in agribusiness produce (like BSE, swine fever and foot-and-mouth disease) has risen. As a result quality assurance plays a major role in agriculture and food processing nowadays. Systems like the ISO-9000 series are implemented, HACCP is obligatory in the meat-processing sector and even at farm level processing industries are forcing quality systems on producers.

From this point of view, profit as a single performance criterion lacks managerial significance (Noori and Radford 1995).

Thirdly, historical data are commonly used in published reports. More emphasis should be placed on cash flows, rather than on past-period profits (Brealey and Myers 1991).

Finally, the relevance of single-business profitability is questioned, since in a supply chain costs of one firm can be revenue for another, and vice versa. Costs as well as risks can easily be transferred from one single firm to another. There is serious danger of opportunistic behaviour if information is not evenly distributed. Stakeholders can be misled; especially the consumers at the end of the supply chain (Barfield et al. 1994). The management of inter-firm interaction and business relationships (by means of legal structures, contracts, covenants, etc.) on the input and output side of corporations is of eminent importance for the survival of companies in food supply chains.

We argue that the measurement of the profitability of the individual firms in a supply chain is inadequate to get insight into the functioning and performance of the channel as a whole. From the previous remarks it follows that performance assessment should be a multidimensional, anticipative and (since individual optimization can hamper overall supply-chain performance) integrated activity.

Accounting principles

The accounting principles that have been developed in the past refer mostly to the information processing of the single (isolated) firm, and are proposed by rule-making bodies, like the EU (4th Directive), IAS (IRFS guidelines), SEC (NYSE prescriptions), GRI (Global Reporting Initiative, stressing sustainability reporting), and national legal authorities. Figure 1 depicts the GRI standards as an example.

There is an exception to the single-firm focus of modern accounting. If companies are economically interwoven (can be considered a 'group'), then also a consolidated financial statement is published. Coordination takes place through a company that, most of the time on the basis of property of share capital, takes the lead over a group of companies and consolidates by integrating assets, liabilities, profits and losses. In a supply chain with (on the basis of share capital rights) a central governing firm, a consolidated report covers the supply chain as a whole.

For other types of governance configurations in supply chains, supply-chain accounting is non-existent. Three types of supply-chain governance can be discerned: the centralized supply chain, the cooperative supply chain and the decentralized (atomic) supply chain. In a centralized supply chain, all information about the strategic and operational issues is concentrated at one of the firms within the channel. Moreover, the central governing firm not only controls the operational processes and strategy, but also the information flows to the contributing firms. The profit from supply-chain activities is captured by this central governing firm, because of its dominant position and power. An example of this governance system is the EUREP-GAP quality system that supermarkets have imposed on the horticultural and agricultural producers.

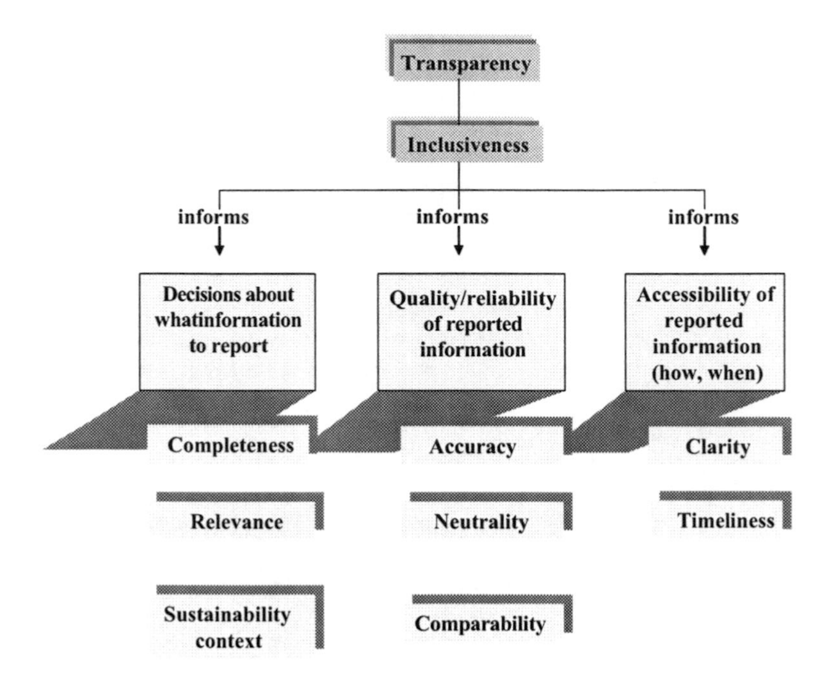

Figure 1. *Reporting principles (source: Global Reporting Initiative 2002, p. 23)*

In a cooperative supply chain, the partners have an equal (or better 'equivalent') say and have the right to be informed on overall chain performance. Surpluses are distributed to the partners in concordance with their contribution to chain efforts. The cooperative supply-chain governance system is the ideal image of a well-functioning, sustainable supply chain, because centralized governance (hierarchy) could just as well take place within the individual firm. On the other end of the continuum, the atomic supply chain uses market price as the only coordination mechanism. Under these rules, accounting is dispersed over different firms and a collective information system is non-existent. Only in the cooperative supply chain, a collectively designed and controlled information system is viable.

Especially in cooperative supply chains that are not governed by decisive (legal) property rights, an integrated or supplementary reporting structure will be necessary, since intellectual, social and physical assets are for a large part not controlled by a single firm, but are common property. Although such a supply-chain governance structure is a channel structure rather than a cooperative in legal sense, its governance resembles the legal cooperative governance in many aspects. Analogous problems are present: performance measurement, redistribution of benefits and joint governance (member influence). Effective vertical coordination and policy disclosure make the development of special standards for supply-chain accounting necessary.

ACCOUNTING PRINCIPLES IN THE COOPERATIVE SUPPLY CHAIN

In this section, we formulate three accounting principles that are of importance in a cooperative supply chain: information availability, asset management and portfolio management.

Chain-accounting standard 1: information availability

The first principle we propose is the reciprocity in information availability: those that supply (high-quality) data, should be able to retrieve (high-quality) data. This principle states the right of chain partners to be informed about chain operations, strategy and outcomes and the obligation to provide information to the system. For instance, processors in the supply chain who contribute to the tracking and tracing system (T&T) should, as compensation, be able to assess how other partners in the chain operate. The supply-chain partners should eventually be provided with information on emergency call-backs of products, but also for instance with information on what consumer groups buy the product and on the level of perceived consumer satisfaction. So the traditional quality-of-information criteria (accuracy, precision, completeness, timeliness, etc.) (Merchant 1998) should be supplemented with reciprocity in information access as an objective and measurable relational criteria (Figure 2).

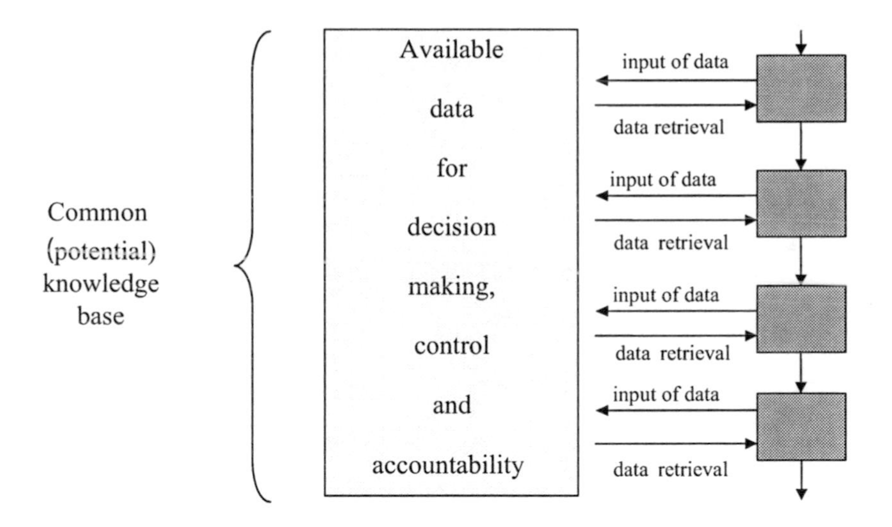

Figure 2. *Reciprocity in information availability*

In a cooperative supply chain, trust should be replaced by (or at least supplemented with) transparency, to reduce information asymmetry. Transparency can be created independently of the established relational quality of supply-chain partners. It can be measured objectively, whereas trust can only be measured by means of interpersonal and subjective relational parameters. In a relationships based on trust, control is purely interpersonal and often 'not done'. Then again, major financial distress cases like Enron, Parmalat or the financial problems of Dutch companies like Ahold and Laurus, would have been prevented if the stakeholder groups would have asked for transparency instead of trusting the managerial decision making beforehand.

Chain-accounting standard 2: asset management

The second chain accounting standard we propose refers to management of assets. In a food supply chain, different partners will invest in systems for common purpose (like T&T, ISO, ECR, etc.). From Williamson's perspective (1983), asset-specific investments occur. Investments in assets that cannot be withdrawn or given another destination without hampering the profitability of individual firms, and eventually of the supply chain as a whole. Common assets (and connected liabilities) eventually produce excess cash flows. If profit measurement is carried out in an economically sensible way, the assets' value represents the present value of the future cash inflows they generate. As was already argued, the traditional accounting value provides a bad representation of the economic significance of (supply-chain) assets.
The specificity of assets has (mutual) dependence as a consequence. With dependence and unevenly distributed information, trust can replace suspicion on the fair redistribution of excess cash flows. Trust itself can be replaced by (or at least supplemented with) transparency of the redistribution of surpluses that is applied.

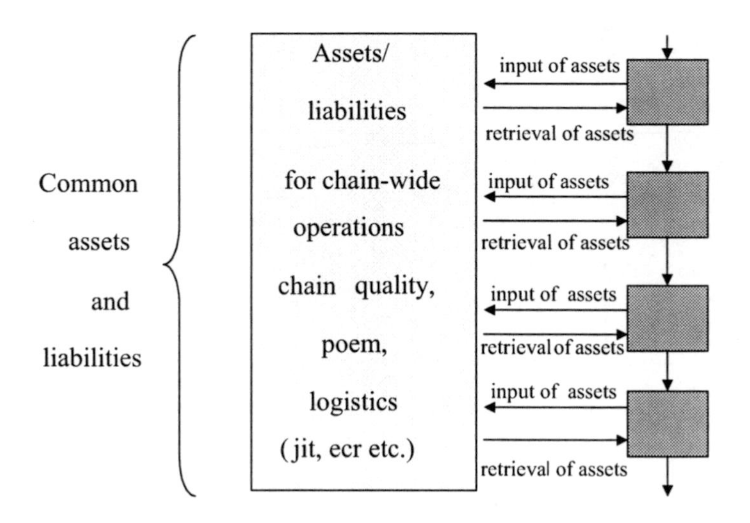

Figure 3. *Asset management*

The following chain-accounting standard can be formulated: the provision of assets should be matched with an equivalent amount of assets (cash flow) in return (see Figure 3).

Determine what is 'equivalent' is a complicated issue in this matter. We already indicated that the solution of relational exchange is rejected. From the perspective of relational exchange, trustworthypartners come up with a mutually beneficial and acceptable long-term agreement themselves on the redistribution of excess cash flows. Since trust issues arise here as well, a redistribution system that is quasi-objective is preferred: the market. Pricing of intermediary products in a supply chain can take place at full costs, full costs plus a profit mark-up, at market prices or at marginal costs. The basic pricing rule for intermediate products however is marginal costs plus opportunity costs (also known as Solomons' rule). Under perfect market conditions, the opportunity costs are the difference between the market price and marginal costs (variable costs in proportional situations). Logically, under these circumstances the market price represents the value of the intermediate product.

But with the absence of a market and the existence of a bottleneck in resource availability, the intermediary product price will be the marginal (= variable) costs plus the shadow price of the capacity employed. This could mean that resource shadow prices become zero (which is the case in situations of homogeneous production and abundant supply; not an unfamiliar situation in agriculture). In the latter case, a cooperative supply chain would allow for chain orchestration to guarantee the continuity of all supply-chain processes. The chain-orchestrating firm or body can only safeguard its own continuity, if at least the full costs of production of the marginal companies are paid. In a cooperative supply chain with mutually dependent relations, the long-term continuity of product procurement should be the leading principle for compensation of the supply-chain partners.

Chain-accounting standard 3: portfolio management

The final accounting standard we wish to propose refers to management of the portfolio of risky projects. A portfolio of supply-chain projects not only creates opportunities for the firms that are involved, but also a diversity of risks. Returns and risks should be in equilibrium: the bigger the opportunities of the single firm, the bigger the contribution to be displayed with respect to risk taking and risk management (Figure 4).

If the market (consumer purchasing power and preferences) is the limiting factor for the level of activity (and profitability) of the supply chain, the retail-companies that control the access to the consumer (and collect the largest part of the opportunities) should be the largest contributor with respect to risk absorption and management. EU legislation is in concordance with this rule: in general, product liability risks downstream the supply chain are bigger than risks upstream. An efficient risk-return trade-off will only take place in markets that operate efficiently. For supply-chain governance this implicates that transparency with respect to supply-chain opportunities and threats should be established. And this exactly is the main task of a viable supply-chain accounting system.

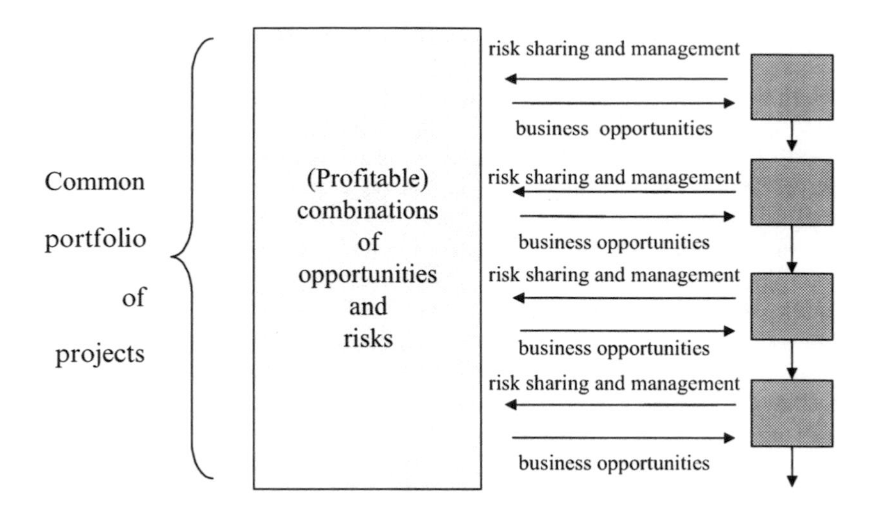

Figure 4. *Portfolio management*

INFORMATION PROCESSING AND INNOVATION

Transparency within a supply chain can be brought about using different instruments. The instruments that are available can be classified as personal, social, cultural, technical and legal. In this paper we concentrate on the technical and legal devices.

An empirical study in 2002, performed by Bremmers et al. (2003), focused on the exploration of innovative legal and technical, i.e. informational, devices, to increase the efficiency and effectiveness of data transfer with respect to environmental performance. They sent questionnaires to 2620 companies in the agri-food industry. For the purpose of this paper we selected 92 companies that had a turnover in 2001 of € 50 mln or more. The following legal technical devices were evaluated:

- integrated environmental permit for the supply chain as a whole;
- integrated environmental report for the supply chain as a whole;
- electronic environmental reporting.

An environmental permit for the supply chain as a whole was considered a big to tremendous improvement by about 20% of the respondents (N = 92). An integrated environmental report was perceived as a big to tremendous amount of improvement by about 32% of the companies. Electronic environmental reporting was considered a big to tremendous improvement by 39.8% of the companies that were questioned.

This result is favourable for the development of electronic environmental reporting (which was implemented in The Netherlands in 2004), but we consider it as only one step towards supply-chain transparency. The reason for this is that environmental disclosure reporting still focuses at the single firm and is limited to environmental performance.

CONCLUSIONS AND SUMMARY

For accounting in supply chains to be effective and efficient, not only technical conditions, like electronic reporting devices, are a prerequisite, but also the availability of (normative) accounting standards. Like all accounting standards, supply-chain accounting standards should mainly come from practice, as well as from theoretical reasoning. In this paper we proposed three accounting standards for cooperative supply chains:

- Reciprocity in information access: those that deliver information to the system should be able to retrieve an equivalent amount. In contrast with centralized and atomic supply chains (in which information is centralized or dispersed, respectively), transparency is best served in cooperative supply chains; they render open access to information concerning supply-chain operations, strategy and results.
- Equivalent cash flows: provision of assets (supply-chain investments) should be matched with an equivalent amount of assets (cash flows) in return.
- Matching risks and returns: the bigger the opportunities for individual firms, the bigger the contribution should be in risk sharing and risk management.

These three rules of reciprocity cause and are an effect of creating a transparent supply-chain policy and performance measurement. Transparency will replace (or at least will have to supplement) *trust* as a measure for relational quality. These three standards are not meant to be exhaustive; other standards should still be developed. We assessed the willingness of system innovations (integrated environmental permit and/or report as well as electronic reporting) for improving the transparency and cooperativeness in supply chains. We found that the devices we proposed are supported by a considerable number of companies. Not only do they improve transparency and cooperative decision making, but they will, if adopted, also reduce the administrative burden of the companies cooperating in supply chains.

REFERENCES

Acs, Z.J. and Gerlowski, D.A., 1996. *Managerial economics and organization*. Prentice Hall, New York.

Alter, S. and Roche, M., 1999. *Information systems: a management perspective*. Addison-Wesley, Reading. 1e uitg.: 1992.

Barfield, J.T., Raiborn, C.A. and Kinney, M.R., 1994. *Cost accounting: traditions and innovations*. 2nd edn. West Publishing, St. Paul.

Beamon, B.M., 1999. Designing the green supply chain. *Logistics Information Management,* 12 (4), 332-342. [http://faculty.washington.edu/benita/paper11.pdf].

Brealey, R.A. and Myers, S.C., 1991. *Principles of corporate finance*. 4th edn. McGraw Hill, New York.

Bremmers, H.J., 1995. *Milieuschade en financieel verslag: de verwerking van milieuschade en -schaderisico's in het externe financieel verslag, met bijzondere aandacht voor de jaarrekening*. Kluwer, Deventer.

Bremmers, H.J., 2000. *Milieuverslaggeving*. Elsevier Bedrijfsinformatie, The Hague.

Bremmers, H.J., 2001. *Towards a new information system for farm management: changing the accounting system for better environmental reporting: paper presented at IFMA conference, 12th July 2001, Papendal, The Neterlands.* IFMA. [http://www.ifma.nl/files/papersandposters/PDF/Papers/Bremmers.pdf].

Bremmers, H.J., Hagelaar, G. and De Regt, M.C., 1996. Bedrijfssituatie en milieuzorg. *Milieu,* 11, 12-19.

Bremmers, H.J., Omta, S.W.F. and Smit, M.E., 2003. *Managing environmental information flows in food and agribusiness chains.* Wageningen UR, Wageningen.

Buzzelli, D.T., 1991. Time to structure an environmental policy strategy. *Journal of Business Strategy,* 12 (2), 17-20.

Carroll, A.B., 1979. A three-dimensional conceptual model of corporate performance. *Academy of Management Review,* 4 (4), 497-505.

Choi, F.D.S., Frost, C.A. and Meek, G.K., 2002. *International accounting.* Prentice Hall, Upper Saddle River.

Clarkson, M.B.E., 1995. A stakeholder framework for analyzing and evaluating corporate social performance. *Academy of Management Review,* 20 (1), 92-117.

Drury, C., 1992. *Management and cost accounting.* 3rd edn. Chapman & Hall, London.

Fowler, S.W., Lawrence, T.B. and Morse, E.A., 2004. Virtually embedded ties. *Journal of Management,* 30 (5), 647-666.

Friedman, M., 1970. The social responsibility of business is to increase its profits [as cited in: Clarkson (1995)]. *The New York Times Magazine* (Sept. 13).

Global Reporting Initiative (ed.) 2002. *GRI guidelines.* GRI, Amsterdam. [http://www.globalreporting.org/guidelines/2002/b23.asp]

Hardaker, J.B., Huirne, R.B.M. and Anderson, J.R., 1997. *Coping with risk in agriculture.* CAB International, Wallingford.

Hart, S.L., 1995. A natural-resource-based view of the firm. *Academy of Management Review,* 20 (4), 986-1014.

Husted, B.W., 2003. Governance choices for corporate social responsibility: to contribute, collaborate or internalize? *Long Range Planning,* 36 (5), 481-498.

Madsen, H. and Ulhøi, J.P., 2001. Integrating environmental and stakeholder management. *Business Strategy and the Environment,* 10, 77-88.

Merchant, K.A., 1998. *Modern management control systems: text and cases.* Prentice Hall, Upper Saddle River.

Noori, H. and Radford, R., 1995. *Production and operations management: total quality and responsiveness.* McGraw-Hill, New York.

Sporleder, T.L. and Peterson, H.C., 2003. Intellectual capital, learning and knowledge management in agrifood supply chains. *Journal on Chain and Network Science,* 3 (2), 75-80.

Tyteca, D., Carlens, J., Berkhout, F., et al. 2002. Corporate environmental performance evaluation: evidence from the MEPI project. *Business Strategy and the Environment,* 11, 1-13.

Waddock, S.A. and Graves, S.B., 1997. The corporate social performance-financial performance link. *Strategic Management Journal,* 18 (4), 303-319.

Wartick, S.L. and Cochran, P.L., 1985. The evolution of the corporate social performance model. *Academy of Management Review,* 10 (4), 758-769.

Welford, R. and Gouldson, A., 1993. *Environmental management and business strategy.* Pitman, London.

Williamson, O.E., 1983. *Markets and hierarchies: analysis and antitrust implications: a study in the economics of internal organization.* Free Press, New York.

Wubben, E.F.M. and Bremmers, H.J., 2003. *Transparency in the Dutch food and agribusiness sector, determining value drivers for (social) responsible corporate reporting.* Wageningen UR, Wageningen.

CHAPTER 4

PRICING AND PERFORMANCE IN AGRI-FOOD SUPPLY CHAINS

FRANK BUNTE

LEI, Wageningen University and Research Centre, P.O. Box 29703, 2502 LS Den Haag, The Netherlands. E-mail: frankbunte@wur.nl

Abstract. This paper explores whether European food-processing and retail industries exert market power towards farmers and consumers. More in particular, this paper analyses (1) whether price changes at the farm level are fully and instantaneously transmitted into changes at the consumer level; and (2) whether there have been changes in the price risk distribution in post-war agri-food supply chains. With respect to the first research question, we do not observe a general pattern of price asymmetry to the disadvantage of farmers and consumers. In general, price symmetry and price levelling are as prevalent as price asymmetry is. With respect to the second question, I observe a shift in price risk from farmers to marketing organizations in the Dutch ware-potato supply chain.
Keywords: supply-chain performance; price transmission; risk distribution; agriculture

INTRODUCTION

In the period between October 2000 and April 2001 there was a dramatic 35% decline in the farmer price for beef in The Netherlands. The price decrease was not followed by subsequent decreases at the wholesale and retail levels. On the contrary, while the wholesale price remained stable, the retail price even rose by 4% (CBS Statline). This difference in price development gave rise to a public debate on price formation in The Netherlands. Price changes at one stage in the food chain are not necessarily transmitted to other stages. Farmer and consumer associations accuse food-processing and retail companies of abusing their market power to increase profit margins. Farmers consequently receive too little and consumers pay too much.

This paper relates industry and supply-chain performance in agri-food supply chains to pricing in agri-food chains, more in particular to price transmission. The paper focuses on one particular aspect of performance: equity. Are the costs and benefits of the production and distribution of food evenly distributed? An uneven distribution of costs and benefits may have consequences for the viability of agri-food chains, since the uneven distribution of costs and benefits may hinder

C.J.M. Ondersteijn et al. (eds.), Quantifying the agri-food supply chain, 39-47.

modernization efforts in agriculture. The paper uses the results of recent empirical studies to study industry and supply-chain performance on pricing.

The paper is constructed as follows. Section 1 presents some statistical artefacts on pricing in agri-food chains. Section 2 defines market performance and pricing in agri-food supply chains. Section 3 defines price transmission and reviews some major empirical studies. Section 4 analyses risk sharing in the Dutch ware-potato supply chain. The paper ends with a brief conclusion.

FARMER'S SHARE IN CONSUMER EXPENDITURE

The farmer's share in consumer expenditure exhibits a steady downward trend in the long run. Figure 1 illustrates that more and more value added is generated in food processing, food trade and food service rather than in agricultural production, especially since the beginning of the 1990s. Interest groups, politicians and media express concern over the decline in the farmer's share in the supply chain's income. The producer's performance is thought to deteriorate with the fall in the farmer's share in value added. However, producer performance is not directly related to his share in value added, but rather to the return on his investment and his labour input. In theory, the producer's share may fall without harming his return on investment and labour, e.g. due to productivity increases. It is also possible that the producer's share remains equal, while the return on his investment and labour deteriorates, e.g. due to cost increases or price squeezes throughout the supply chain invoked by retailer price competition. Shares in value added provide useful information on supply chain performance, but not enough information to evaluate farmer performance.

Figure 1 illustrates that the farmer's share in consumer expenditure on food has been falling. Are there any reasons to be concerned about this fall? In order to answer this question, we briefly discuss the main reasons behind the long-term fall in the farmer's share in consumer expenditure on food (De Bont et al. 2000).

Consumption patterns

There is an important shift in food consumption from fresh produce to processed produce and from home consumption to out-of-home consumption. Processed food involves more value added than fresh produce, and out-of-home consumption involves more value added than home consumption. The shift in consumption patterns is directed to food products in which the processing and distribution trades have a larger share (Figure 1).

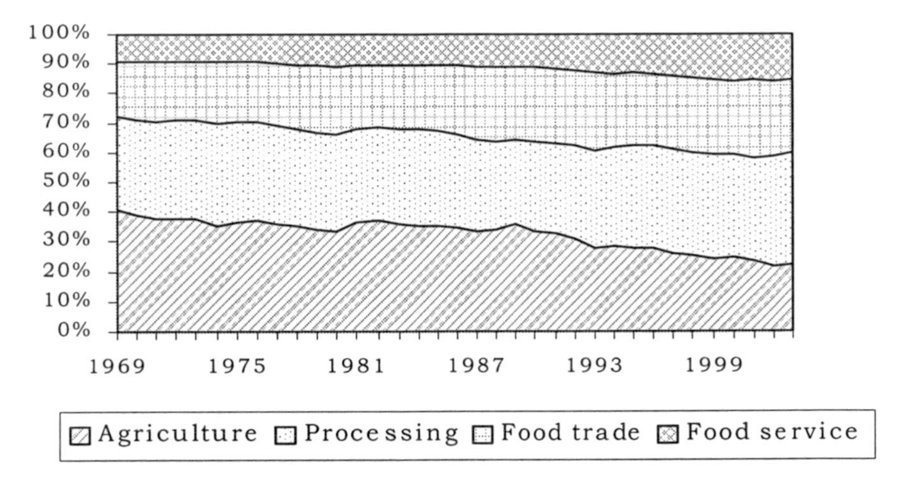

Figure 1. Shares in value added from food production and distribution: The Netherlands, 1969-2003 (source: CBS Nationale Rekeningen)

Productivity patterns

Factor productivity increases faster in agriculture than in manufacturing and services. Bernard and Jones (1996) indicate that factor productivity in agriculture increases at a 40% and 180% higher rate than factor productivity in manufacturing and services, respectively. As a result, agriculture employs fewer production factors and faces a drop in its share in value added of food products.

Market power

Farmers' share in consumer expenditure may also decrease due to abuse of market power by the processing and distribution trades. Downstream industries may extract either lower purchase prices or higher customer prices (or both). If the processing and distribution trades are able to exploit market power, they increase the wedge between consumer and farmer prices and reduce farmers' share of consumer expenditure. Up to now, the empirical literature has found limited evidence for abuse of market power in food processing and distribution (Peltzman 2000).

The change in consumption patterns and the difference in productivity increases explain the long-term gradual decline in farmers' share in consumer expenditure on food. The change in consumption patterns and the difference in productivity increases involve structural factors underlying consumer demand and industry costs. Market power gives an additional ground for developments from and shifts around the structural trend in the distribution of expenditure shares. Market power is a major policy concern, since it influences supply-chain performance.

EFFICIENCY, EQUITY AND PRICING

Welfare theory may be used to assess industry and supply-chain performance on the basis of measures of social welfare. Social welfare depends on the welfare of society's main stakeholders: producers, consumers and taxpayers-citizens. Social welfare – in this case social supply-chain performance – depends on two elements: (1) efficiency (profit) and (2) equity (people). Efficiency is concerned with the creation of value added; equity is concerned with the division of value added over the respective stakeholders.

Efficiency and equity are not necessarily compatible. Efficient solutions may be very 'unequitable'. Maximizing value added is not necessarily beneficial to all stakeholders concerned. Tirole (1988) evaluates supply chain coordination devices by assessing their impact on supply chain performance, more in particular producer surpluses throughout the supply chain. Many solutions Tirole (1988) suggests involve monopoly solutions. These solutions maximize the supply chain's value added. However, value added accrues to the monopolist only, either a processor or a retailer. All other parties do not necessarily gain from supply-chain efficiency. Monopoly profits may be redistributed but there is no reason why they should be. Monopoly power – or more generally market power – does not disappear with supply-chain coordination. Supply-chain coordination may even be a mechanism to establish and maintain market power. So we conclude that supply-chain performance is more than efficiency and that equity matters. The rest of this paper, focuses on equity.

Pricing in supply chains is highly relevant to assess efficiency and equity in supply chains. Clarke et al. (2002) distinguish five aspects when discussing pricing in supply chains:

1. Price levels and profit margins. Firms may earn 'excess' profit margins. Profits are considered to be excessive if they exceed the level deemed necessary to induce firms to produce, to invest and to innovate. Firms may also earn insufficient profits. Profits are insufficient in the sense that they are not high enough to induce firms to produce, to invest and to innovate.
2. Price changes. Buying firms may or may not react to changes in supplier prices (or final prices). Firms may react instantaneously or with a lag, and they may react asymmetrically to decreases and increases in supplier prices. Asymmetries in the reaction to supplier prices generate temporary profits.
3. Price structure. There is more to pricing than unit prices. Firms may also agree to fixed payments, e.g. listing fees, slotting allowances and even retrospective payments.
4. Non-price aspects. Contracts also lay down product specifications. These specifications may substitute for price and other financial transfer clauses.
5. Price risk. A firm's well-being does not only depend on expected income, but also on the price and income risks the firms are exposed to. Price risks make firms more vulnerable, *ceteris paribus*.

The equity issue is at stake when firms are able to exert market power. Market power may be exerted on a permanent or temporary basis, by charging high consumer prices and by commanding low supplier prices (points 1 and 2), by extracting fixed payments (point 3), by enforcing non-price specifications (point 4) and by shifting price risks to other supply-chain parties (point 5).

Ideally, empirical research into supply-chain pricing involves all five elements. However, due to restrictions in time, data and money, empirical research is confined to some research questions. These research questions typically have a partial nature and are restricted to areas for which data are available. This explains why one may indicate several white spots in empirical research. Research establishing the return on investment of subsequent links in supply chains is scarce (point 1). Empirical research on price transmission is abundant (point 2). We delve into this issue in the next section. Empirical attention for financial conditions other than unit prices is new (point 3). Systematic knowledge is not available yet. Empirical attention for the financial consequences of non-price specifications is also new (point 4). Analyses of price risks in agriculture are abundant (point 5), but generally do not address risk sharing in agri-food supply chains, especially not in developed countries. In general, there is little empirical research explaining the price patterns found.

PRICE TRANSMISSION

Price transmission is one of the most heavily studied equity issues related to pricing studied in Industrial Organization. An important part of the empirical applications refer to agri-food chains. Price transmission refers to the way prices at one level in the product chain react to changes at another level. Market power may explain that price changes at one level are not transmitted to other levels. There are three types of imperfect price transmission:

1. Price changes are not fully transmitted.
2. There is a time lag between the price adjustments at the respective stages.
3. There is an asymmetry in reaction between positive and negative price shocks.

Imperfections in price transmission may be due to, among other things, market power or adjustment costs. Market power may explain why prices are not fully transmitted. Oligopolistic and oligopsonistic interdependence may give rise to lags in price adjustment. The risk of invoking a price war may make firms reluctant to lower prices. This may cause an asymmetry in the price reaction to positive versus negative price shocks.

Due to several adjustment costs (labelling, advertising and goodwill) changing prices may be expensive. Adjustment costs thus give rise to reaction lags. In combination with other arguments, such as inflation (Ball and Mankiw 1994), stock building (Blinder 1982) and perishability (Ward 1982), adjustment costs may also cause price asymmetries. Adjustment costs thus give rise to price levelling. The marketing literature gives several other arguments for this phenomenon. Apart from market power and adjustment costs, non-linearities in demand and supply may give rise to imperfections in price transmission.

Table 1. *Results of price-asymmetry studies (source: Meyer and Von Cramon-Taubadel 2002)*

	Test method					
	All methods	First differences	Summation first differences	Error correction	Threshold methods	Other methods
Number of tests	197	93	47	31	10	18
Symmetry	102	30	36	17	2	17
Asymmetry	95	63	11	14	8	1
Asymmetry (%)	48	68	23	45	80	6

There is a wide body of empirical literature on asymmetric price transmission. Meyer and Von Cramon-Taubadel (2002) have summarized the results of 38 studies, 25 of which refer to agricultural products. In these studies, 197 estimations have been performed. These estimations are based on different methods, among other things because estimation methods have been improved through time. Table 1 summarizes the estimation results. Table 1 shows that price asymmetry is a recurrent phenomenon. Almost 50% of the studies found price asymmetry. Note, however, that the estimation results seem to depend on the estimation method employed. Peltzman (2000) also establishes asymmetry in two thirds of the 242 product chains analysed[1].

Recently, London Economics (2004) studied price transmission in European agri-food supply chains in the 1990s (Table 2). Table 2 indicates whether prices are transmitted symmetrically (green), asymmetrically (red) or levelled off (yellow). Table 2 analyses the transmission of price shocks both from upstream to downstream (U-D, from farmer to retailer) and the other way round (D-U, from retailer to farmer). London Economics establishes asymmetry in 13 out of 82 cases. Price symmetry (46 cases) and price levelling (23 cases) are more prevalent. London Economics concludes that there is no general pattern of price asymmetry in agri-food chains.

We conclude as follows. Price asymmetry is a recurrent phenomenon, in supply chains in general and in agri-food supply chains in particular (Meyer and Von Cramon-Taubadel 2002; London Economics 2004). However, there is no general pattern of price asymmetry in agri-food supply chains. Since there is no general pattern of price transmission, a general explanation that pertains to all supply chains cannot be drawn. This holds for both retail concentration and menu costs. The empirical literature still has problems explaining the price patterns found.

***Table 2**. Price symmetry and asymmetry in European supply chains (source: London Economics 2004)*

	Austria		Denmark		France		Germany		Ireland		Nether-lands		Spain		UK	
	U-D	D-U	U-D	D-U	U-D	D-U	U-D	D-U	U-D	D-U	U-D	D-U	U-D	D-U	U-D	D-U
Apples	Price symmetry	Price symmetry					Price symmetry	Price symmetry							Price symmetry	Price symmetry
Beef					Price levelling	Price levelling			Price levelling	Price levelling	Price symmetry	Price asymmetry			Price symmetry	Price symmetry
Bread			Price symmetry	Price levelling	Price asymmetry	Price levelling					Price symmetry	Price symmetry			Price symmetry	Price asymmetry
Butter			Price levelling	Price levelling	Price levelling	Price asymmetry	Price symmetry	Price symmetry							Price levelling	Price asymmetry
Carrots	Price symmetry	Price symmetry					Price symmetry	Price asymmetry							Price symmetry	Price symmetry
Cheese			Price levelling	Price levelling	Price levelling	Price asymmetry	Price symmetry	Price asymmetry							Price levelling	Price levelling
Chicken					Price symmetry	Price symmetry	Price asymmetry	Price symmetry								
Eggs			Price symmetry	Price symmetry	Price levelling	Price symmetry					Price symmetry	Price symmetry			Price symmetry	Price symmetry
Flour			Price symmetry	Price symmetry											Price levelling	Price levelling
Lamb					Price levelling	Price levelling									Price asymmetry	Price symmetry
Milk			Price levelling	Price levelling	Price levelling	Price asymmetry	Price asymmetry	Price symmetry							Price symmetry	Price asymmetry
Potato			Price symmetry	Price symmetry			Price symmetry	Price symmetry			Price symmetry	Price symmetry	Price symmetry	Price symmetry	Price symmetry	Price symmetry

Legend: Price symmetry (diagonal hatching); Price asymmetry (horizontal hatching); Price levelling (vertical hatching)

This section presents the results of empirical research into price risk distribution. This section indicates that there is evidence of a 'power shift' in the Dutch ware-potato supply chain from farmers to wholesale and retail traders.

Marketing organizations and potato farmers engage in contracts with fixed and variable rewards. Marketing organizations and farmers have a principal–agent relation in which work effort and income risk are the main arguments. Marketing organizations are primarily interested in promoting farmer work effort at the lowest possible cost. Work effort may be enhanced by profit sharing, i.e. by a variable reward. Farmers are not only interested in maximizing expected income, but also in income insurance. Farmer income typically depends on a few products – often even one product – while marketing organizations are well able to diversify their product portfolio. Due to the associated difference in income risk as well as to a difference in attitude towards risk, marketing organizations tend to insure farmers against income variability. The fixed reward creates income certainty for the farmer.

Consequently, there is a trade-off between both arguments in the principal–agent relation: efficiency (variable reward) and insurance (fixed reward). Given further specifications, the optimum trade-off may be derived theoretically and the actual trade-off may be determined empirically.

Kuwornu et al. (2004) estimate the development of price and income risks for both farmers and marketing firms in the Dutch ware-potato supply chain. The price and income risks potato farmers bear have steadily increased in the period between 1946 and 1996, especially since 1975 (Figure 2). The price and income risks of marketing firms have diminished in the same period and have become minimal since 1985. Marketing firms have shifted price and income risks to farmers. This fact may be explained by a decrease in farmer risk aversion in the 1990s (Kuwornu et al. 2004). However, the results indicate that, while farmers still demand risk insurance whereas marketing firms do not, farmers actually insure marketing firms against price and income risks. The change in price risk distribution may be due to a shift in

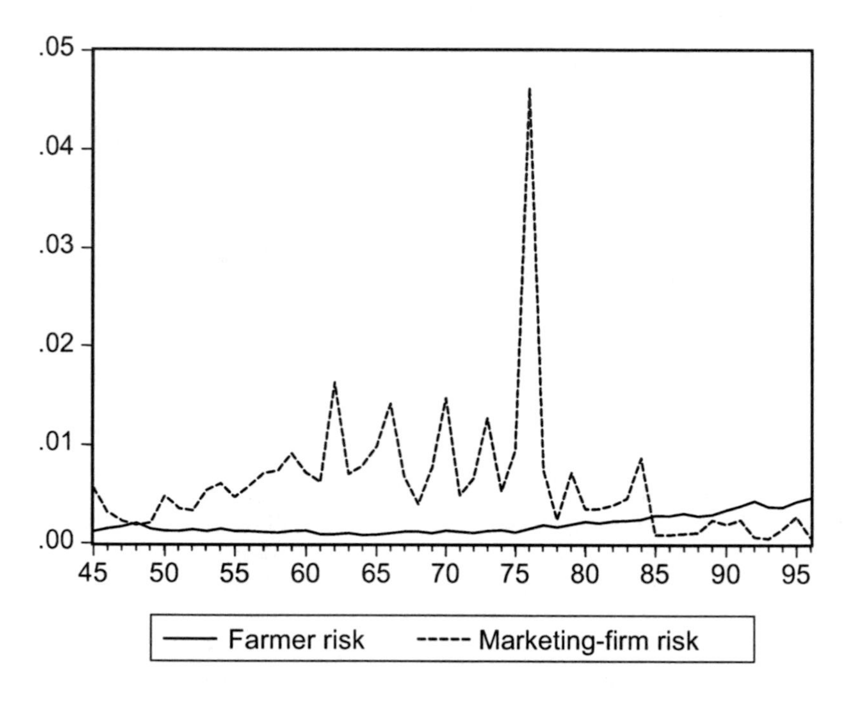

Figure 2. *Farmer and marketing-firm income risks in the Dutch ware-potato supply chain (billion €; source: Kuwornu et al. 2004)*

bargaining power from farmers to retailers. The change in the supply–demand relations after the second world war and the rise in wholesale and retail concentration may have led to a shift in income risk at the expense of farmers. The change in price risk distribution may also be due to a change in supply-chain efficiency requirements. Farmers may be given more price incentives in order to enhance supply-chain value added.

CONCLUSION

This paper finds limited evidence for the abuse of market power in the European food processing and retail trade. There is no general pattern of price asymmetry in European agri-food supply chains. However, for ware potatoes there is evidence of a shift in income risk from marketing organizations to farmers. The paper also indicates that performance evaluation in agri-food supply chains deserves further attention. This holds notably for analysis into the return to investments of subsequent links in the supply chain and non-traditional financial transactions, such as slotting allowances. However, the biggest challenge that lies ahead is explaining what factors contribute to good performance and what factors do not. Measuring performance is one thing, explaining it is another.

NOTES

[1] Peltzman's results are not summarised by Meyer and Von Cramon-Taubadel (2002)

REFERENCES

Ball, L. and Mankiw, N.G., 1994. Asymmetric price adjustment and economic fluctuations. *Economic Journal,* 104, 247-261.

Bernard, A.B. and Jones, C.I., 1996. Productivity across industries and countries: time series theory and evidence. *Review of Economics and Statistics,* 78 (1), 135-146.

Blinder, A.S., 1982. Inventories and sticky prices: more on microfoundations of macroeconomics. *American Economic Review,* 72 (3), 334-348.

Clarke, R., 2002. *Buyer power and competition in European food retailing*. Edward Elgar, Cheltenham.

De Bont, C.J.A.M., Bolhuis, J., Bunte, F.H.J., et al., 2000. *Prijzenswaardig: prijzen en prijsopbouw in de agrokolom.* Lei, Den Haag. [http://www.lei.wageningen-ur.nl/publicaties/PDF/2000/3_xxx/3_00_01.pdf]

Kuwornu, J.K.M., Kuiper, W.E. and Pennings, J.M.E., 2004. Time series analysis of a principal-agent model to assess risk shifting in agricultural marketing channels: an application to the Dutch ware potato marketing chain. *In:* Van Huylenbroeck, G., Lauwers, L. and Verbeke, W. eds. *Role of institutions in rural policies and agricultural markets*. Elsevier, Amsterdam, 255-271.

London Economics, 2004. *Investigation of the determinants of farm-retail price spreads: final report to DEFRA*. London Economics, London. [http://statistics.defra.gov.uk/esg/reports/pricespreads/wholerep.pdf]

Meyer, J. and Von Cramon-Taubadel, S., 2002. *Asymmetric price transmission: a survey: paper at the 10th EAAE conference in Zaragoza*. [http://www.jochenmeyer.de/pdf/meyer-cramon.pdf]

Peltzman, S., 2000. Prices rise faster than they fall. *Journal of Political Economy,* 108 (3), 466-502.

Tirole, J., 1988. *The theory of industrial organization*. MIT Press, Cambridge.

Ward, R.W., 1982. Asymmetry in retail, wholesale, and shipping point pricing for fresh vegetables. *American Journal of Agricultural Economics,* 64 (2), 205-212.

CHAPTER 5

PERFORMANCE INDICATORS IN AGRI-FOOD PRODUCTION CHAINS

LUSINE ARAMYAN[#*], CHRISTIEN ONDERSTEIJN[#], OLAF VAN KOOTEN[##], ALFONS OUDE LANSINK[#]

[#] *Business Economics Group, Wageningen University, P.O. Box 8130, 6700 EW Wageningen, The Netherlands*
[##] *Horticultural Production Chains Group, Wageningen University, Marijkeweg 22, 6709 PG Wageningen, The Netherlands*
**E-mail: Lusine.Aramyan@wur.nl*

Abstract. The last decade has seen an increasing interest in indicators of supply-chain performance. A large number of various performance indicators have been used to characterize supply chains, ranging from highly qualitative indicators like customer or employee satisfaction to quantitative indicators like return on investments. This large number of different performance indicators, and the lack of consensus on what determines performance of supply chains, complicates the selection of performance measures. Furthermore, combining these indicators into one measurement system proves to be difficult. Efforts as well as progress have been made in this area but supply-chain performance measurement received little or no attention in the field of food and agribusiness. This paper provides a literature review on existing performance indicators and models, and discusses their usefulness in agri-food supply chains. Furthermore, based on this overview, a conceptual framework is developed for further research in this area.
Keywords: measure; efficiency; responsiveness; flexibility; food quality; framework

INTRODUCTION

A supply chain is generally defined as a network of physical and decision-making activities connected by material and information flows that cross organizational boundaries (Van der Vorst 2000). According to Lambert and Cooper (2000) there are four main characteristics of a supply chain: first it goes through several stages of increasing intra- and inter-organizational, vertical coordination. Second, it includes many independent firms, suggesting that managerial relationship is essential. Third, a supply chain includes a bi-directional flow of products and information and the managerial and operational activities. Fourth, chain members aim to fulfil the goals to provide high customer value with an optimal use of resources. An agri-food chain

C.J.M. Ondersteijn et al. (eds.), Quantifying the agri-food supply chain, 49-66.

is nothing more than a supply chain which produces and distributes an agricultural or horticultural product and where product flows and information flows take place simultaneously (Bijman 2002).What makes agri-food supply chains different from other supply chains is (1) the nature of production, which is partly based on biological processes, thus increasing variability and risk; (2) the nature of the product, which has specific characteristics like perishability and bulkiness that require a certain type of supply chain; and (3) the societal and consumer attitudes towards issues like food safety, animal welfare and environmental pressure.

Within a chain, coordination may take various forms: vertical integration, long-term contracts or market transactions. Recent studies have shown that in agri-food supply chains, transactions are undergoing several changes (Bijman 2002). Most agri-food sectors are moving closer to vertical coordination. Some industries (e.g. poultry) developed tight vertical coordination some time ago, while in others it is a relatively new phenomenon (Hobbs and Young 2000) The major change is the shift from a production orientation to a market orientation in the strategy of producers. This change leads to an increase in the information exchange among agri-food chain members. Another change relates to product innovation, which has become very important in agri-food chains. All these changes are the result of an increasing consumer demand for more quality and a larger variety of products. Moreover, issues such as food safety and production conditions are major concerns for consumers nowadays. Apart from the changes in preferences of consumers, there are also structural changes in processing and retailing of agri-food products. Processors and retailers have become larger and more internationalized. Agricultural policies have undergone several changes at national and EU level as well, which have led to a decreasing level of market protection and to shifting priorities in spending public funds.

The development of more integrated supply chains was not followed by simultaneous development of supply-chain performance indicators and metrics in order to assess the effectiveness of a particular chain organization (Gunasekaran et al. 2001). This is not only true for agri-food chains, but reflects the general developments in this area. Measurement of supply-chain performance gives decision-makers inside (e.g., producers, distributors, marketers) and outside (e.g., policy-makers, investors) the supply-chain information for decision making, policy development, etc. The goal of this study is to develop a flexible conceptual framework for measuring the performance of agri-food supply chains that can be used by different decision-makers. The objectives of this paper are therefore:

- to provide a literature review on existing performance indicators in supply chains;
- to give an overview of different methods and models used to measure performance of supply chains;
- based on the literature review, to develop a conceptual framework on selection of performance indicators in agri-food supply chains.

PERFORMANCE INDICATORS IN SUPPLY CHAINS

In 1992 Lee and Billington found that no adequate supply-chain metrics exist, and firms, even if they are participating in coordinated supply chains, only aim at achieving their own performance standards. (Beamon 1999) looked at performance indicators used in supply-chain modelling and concluded that "current supply chain performance measurement systems are inadequate because they rely heavily on the use of cost as primary measure, they are not inclusive, they are often inconsistent with the strategic goals of the organization, and do not consider the effects of uncertainty". A few years later, Gunasekaran et al. (2001) reviewed the literature of performance metrics of supply chains again and concluded that there is still a lack of a balanced approach with regards to financial as well as non-financial indicators and the number of performance indicators to be used. Furthermore, no distinction is made between indicators of operational, tactical and strategic level. In their work Gunasekaran et al. (2001) develop a conceptual model for supply-chain performance at three levels: strategic, tactical and operational. There seems to be consensus about the fact that no supply-chain measurement system exists that is inclusive, universal and measurable as well as consistent (Beamon 1998). There is less agreement, however, on the matter of what such a system should look like. Hannus (1991)[1] emphasizes that a supply-chain measurement system should reflect the objectives of main interest groups (customers, owners and personnel), it should combine operational and financial follow-up data, and link operational objectives to critical success factors and goals. He suggests using three main categories of performance indicators: customer satisfaction, flexibility and efficiency, and to pay attention to three main indicators such as quality, time and costs in these main categories. In his paper he developed an approach for business-process re-engineering. This approach was lately described in the work of Korpela et al. (2002) as the basic theoretical framework in supply-chain development and combined with the theory of analytic hierarchy process (AHP). This paper was an attempt to demonstrate how the analytic hierarchy process can be used for supporting the supply-chain development process.

Murphy et al. (1996) conducted a two-stage study, where the first stage gave an overview of performance indicators and their dimensions used in literature from 1987 to 1993 and the second stage examined the relationship between performance variables and the existing performance dimensions. In their work Murphy et al. (1996) used 19 performance indicators, mostly being of financial nature such as net income or return on investments. In 1999, Beamon (1999) suggested a system of three dimensions: resources (i.e., efficiency of operations), output (i.e., high level of customer service) and flexibility (i.e., ability to respond to a changing environment). Persson and Olhager (2002) adhered to this three-dimension system. Based on results of a simulation model they concluded that good quality and short lead-times in integrated and synchronized supply chains lead to superior performance. The pay-off in terms of total cost is more than proportional to the improvements in quality and lead-times.

Li and O'Brien (1999) suggested a model to improve supply-chain efficiency and effectiveness based on four criteria: profit, lead-time performance, delivery promptness and waste elimination. Their model analyses the supply-chain performance at two levels: the chain level and the operational level. At the chain level, assumptions for these four criteria are set for each supply-chain stage so that the supply-chain performance can meet the customer service objectives. At the operations level, manufacturing and logistics procedures are optimized under the given objectives and three different strategies. The results of the model revealed that lead-time performance is the most influential factor for the choice of the strategy. Berry and Naim (1996) and later on Li and O'Brien (1999) emphasize that the efficiency of supply chains can generally be improved by reducing the number of manufacturing stages, reducing lead-times, working interactively rather than independently between stages and speeding up the information flow. Efficiency and effectiveness were also used in the work of Lai et al. (2002) to evaluate the supply-chain performance in transport logistics. Lai et al. identified three dimensions of supply-chain performance in transport logistics. Those dimensions are service effectiveness for shippers, operational efficiency and service effectiveness for consignees. Within these dimensions they identified four performance indicators such as responsiveness, reliability, costs and assets.

Van de Vorst (2000) distinguished several performance indicators for food supply chains on three levels: supply chain, organization and process. At supply-chain level five indicators are distinguished: product availability, quality, responsiveness, delivery reliability and total supply-chain costs. At organization level again five indicators are distinguished: inventory level, throughput time, responsiveness, delivery reliability and total organizational costs. Finally at process level four indicators are distinguished: responsiveness, throughput time, process yield and process costs. Thonemann and Bradley (2002) follow the line of Eppen (1979) and analyse the effect of product variety on supply-chain performance, measured in terms of expected lead-time and expected cost at the retailer level in a single-manufacturer and multiple-retailer model. They showed that underestimating the cost of product variety leads companies to offer product variety that is greater than optimal. The authors also demonstrate how supply-chain performance can be managed by reducing the set-up time, the unit-manufacturing time, the number of retailers or the demand rate.

In 2003 Claro et al. built an integrated framework for Dutch potted-plant and flower production that aimed at the combination of constructs on the transaction, dyadic and business-environment level for testing their impact on relational governance and performance. Each of these three levels consists of different determinants. Determinants of transaction level are exchange mode, human and physical transaction-specific assets, determinants of dyadic level are length of business interaction and organizational trust, and finally, determinants of business-environmental level are network intensity and environmental instability. As an indicator of relational governance they used joint planning and joint problem solving and as indicator of performance they used sales growth rate and perceived satisfaction. The results revealed that the dimensions of relational governance

Table 1. *Literature review on supply-chain performance measures*

Author	Sector	Customer responsiveness	Efficiency	Flexibility	Other	Number of indicators
Eppen (1979)	Steel production		X			1
Hannus (1991)[1]	Manufacturing		X	X		3
Lee and Billington (1992)	Manufacturing		X			1
Berry and Naim (1996)	Manufacturing	X		X	X	4
Murphy et al. (1996)	Different industries		X		X	35
Beamon (1998)	Manufacturing	X	X	X	X	16
Beamon (1999)	Manufacturing	X	X	X	X	33
Li and O'Brien (1999)	Manufacturing	X	X	X	X	11
Talluri et al. (1999)	Manufacturing	X	X		X	9
Van de Vorst (2000)	Food	X	X	X	X	8
Gunasekaran (2001)	Not specified	X	X	X	X	43
Thonemann and Bradley (2002)	Manufacturing	X	X			2
Korpela et al. (2002)	Not specified	X	X	X		3
Lai et al. (2002)	Transport	X	X	X	X	4
Talluri and Baker (2002)	Manufacturing	X	X	X	X	15
Persson and Olhager (2002)	Manufacturing	X	X	X	X	7
Claro et al. (2003)	Horticulture		X	X		2
Gunasekaran (2004)	Different industries	X	X	X	X	45

[1] The work of Hannus (in Finnish) is taken from the paper by Korpela et al. (2002)

positively affect sales growth and perceived satisfaction, except that joint planning is not related to perceived satisfaction.

The literature review shows that many attempts have been made to develop a measurement system for supply chains. None have been successfully incorporated in practice. Table 1 summarizes the papers described above in the most commonly used categorization: efficiency, flexibility and responsiveness. Responsiveness aims at a high level of customer service and may include fill rate, product lateness, customer response time, lead-time and shipping errors. Flexibility indicates the degree to which the supply chain can respond to a changing environment. Flexibility includes customer satisfaction and reductions in the number of backorders, lost sales and late orders. Efficiency aims to maximize value added by the process and minimize the cost absorbed in inventories. It includes several indicators, but the most commonly used are costs, profit, return on investment and inventory (inventory investments, inventory obsolescence).

As can be seen from Table 1 research on agri-food supply chains is rather limited. Furthermore, the literature review showed several performance indicators that could not be placed under one of the three categories and are therefore placed in a category 'other'. These performance indicators are, for instance, range of products and services, variations against budget, product differentiation, stock-out probability, etc.

MODELS AND METHODS TO ASSESS SUPPLY-CHAIN PERFORMANCE

Different methods exist that can incorporate multiple performance indicators into one measurement system. Some of the best-known are the Supply-Chain Council's Supply-Chain Operations Reference (SCOR®) model, the Balanced Scorecard, Multi-Criteria Analysis, Data-Envelopment Analysis, Life-Cycle Analysis, and Activity-Based Costing. The review in this section discusses different measurement methods and the advantages and disadvantages of these methods.

The Supply-Chain Council's SCOR® model is a standard supply-chain process reference model designed to fit all industries (Supply-Chain Council 2004). This model provides guidance on the types of metrics decision-makers can use to develop a balanced approach towards measuring the performance of an overall supply chain. The SCOR® model advocates a set of supply-chain performance indicators as a combination of: 1) reliability measures (e.g., fill rate, perfect order fulfilment); 2) cost measures (e.g., cost of goods sold); 3) responsiveness measures (e.g., order fulfilment lead-time); and 4) asset measures (e.g., inventories). The SCOR® model directly addresses the needs of supply-chain management at the operational level. One of the tenets of the SCOR® model is that a supply chain must be measured and described in multiple dimensions. These dimensions include reliability, responsiveness, flexibility, cost, and efficiency of asset utilization. The SCOR® model is a cross-industry model that decomposes the processes within a supply chain and provides a best-practice view of supply-chain processes The advantages of the SCOR® model are that it takes into account the performance of the overall supply chain; it proposes a balanced approach by describing performance of the supply chain in multiple dimensions. Disadvantages include the fact that SCOR® is very operations-oriented and does not attempt to describe all relevant business processes or activities such as sales and marketing, research and technology developments, product developments and post-delivery customer support. Secondly, and related tot the previous disadvantage, SCOR® assumes but does not explicitly address training, quality, information technology and administration (Supply-Chain Council 2004). Scientific research using the SCOR® model is limited. Based on the SCOR® model (developed by Stephens 2000) Lai et al. (2002) used the model to evaluate supply-chain performance. Lai et al. identified three dimensions of supply-chain performance in transport logistics, which are service effectiveness for shippers, operational efficiency, and service effectiveness for consignees. Based on these three dimensions a 26-item supply-chain performance measurement instrument was constructed, which was tested empirically and found to be reliable and valid for evaluating supply-chain performance in logistics. Wang (2003) related product characteristics to supply-chain strategy in order to analyse a product-driven supply-chain selection, and adopted SCOR® model level-1 performance metrics as the decision criteria for supplier selection. Based on the SCOR® model they developed an analytic hierarchy process (AHP) with overall objective to achieve optimal supplier efficiency. Then, authors developed an integrated multi-criteria decision-making methodology based on AHP and pre-emptive goal programming (PGP) so that it takes into account both qualitative and quantitative factors in supplier selection. They found that integrated AHP-PGP methodology can select the best set of multiple suppliers to satisfy suppliers' capacity constraint.

The Balanced Scorecard is a popular performance measurement scheme initially developed by Kaplan and Norton (1992). This method employs performance metrics from financial (e.g., cost of manufacturing and cost of warehousing), customer (e.g., on-time delivery and order fill rate), business process (e.g., manufacturing adherence-to-plan), innovation and technology perspective (e.g., new-product development cycle time). By combining these different perspectives, the balanced scorecard helps a manager to understand the interrelationships and trade-offs between alternative performance metrics and leads to improved decision making. This method is not specifically designed for supply chains but could be adapted to focus on supply-chain performance. The Balanced Scorecard is more tactical and strategically oriented compared with the SCOR® model, which is an operation-oriented method.

The advantages of the Balanced Scorecard are that it uses four performance dimensions, both financial and non-financial, which ensures that management is given a balanced view on performance. Finally, a top-level strategy and middle-management level actions are clearly connected and appropriately focused. Disadvantages are that this approach requires considerable thoughts and effort to develop an appropriate scorecard, the scorecard does not include market-oriented performance indicators, and complete implementation should be staged (Coronel 1998). The Activity-Based Costing (ABC) method is based on accounting methods and involves breaking down activities into individual tasks or cost drivers, while estimating the resources (i.e., time and costs) needed for each one. Costs are then allocated based on these cost drivers, such as allocating overhead either equally or based on less-relevant cost drivers. This approach allows for better assessing the productivity and costs of a supply-chain process. By means of the ABC method companies can more accurately assess, e.g., the costs of services for a specific customer or the costs of marketing a specific product. Hence, businesses can understand the factors that drive each major activity, the costs of activities, and the relationship between activities and products. ABC analysis does not replace traditional financial accounting, but provides a better understanding of performance by looking at the same numbers in a different way (Lapide 2000).

The advantages of ABC are that it gives more than just financial information and it recognizes the changing cost behaviour of different activities as they grow and mature. Disadvantages are that ABC, like the Balanced Scorecard, is not developed for supply chains but could be adapted. Furthermore, data collection can be costly and time-consuming. While it is difficult to determine appropriate cost drivers in ABC for businesses, this may even prove to be a bigger challenge for supply chains. ABC focuses primarily on costs.

Traditional accounting is focused on short-term financial results like profits and revenues, providing little insight into the success of an enterprise towards generating long-term value to its shareholders. To overcome this problem, the estimation of a company's Economic Value-Added (EVA) was introduced. This method is based on the assumption that shareholder value is increased when a company earns more than its cost of capital. Unlike Balanced Scorecards, which offer a functional focus toward performance, the EVA offers a project focus. EVA attempts to quantify value created by an enterprise, basing it on operating profits in excess of capital employed (through debt and equity financing). EVA metrics are less useful for measuring detailed supply-chain performance. They can be used, however, as the supply-chain metrics within an executive-level performance scorecard, and can be included in other measurement systems such as, e.g., the Logistics Scoreboard approach (Lapide 2000). The advantages of EVA are that it explicitly considers the cost of capital and allows projects to be viewed separately. Disadvantages of EVA are its difficulties with computations and allocation of EVA among divisions.

Multi-Criteria Analysis (MCA) establishes preferences between options by reference to an explicit set of objectives that the decision-maker has identified, and for which he or she has established measurable criteria to assess the extent to which the objectives have been achieved. This method is designed to support decision-makers facing complex, multi-dimensional problems (Romero and Rehman 2003). Several techniques exist, like direct analysis of the performance matrix, multi-attribute utility theory, linear additive models, procedures that use qualitative data inputs and so on. The following steps are carried out by the decision-makers in MCA: 1) identify the feasible alternatives or preferred outcomes; 2) identify the criteria by which to judge these outcomes; 3) apply appropriate weights on each of the criteria that reflect their particular preferences.

One of the biggest advantages of MCA is that it facilitates a participatory approach to decision making. Another advantage is that the interactive nature of the approach enables both analyst and decision-maker to learn more about the problem. Finally, it is suitable for problems where monetary values of the effects are not readily available. On the other hand, although MCA does not necessarily require quantitative or monetary data, the information requirements to derive the weights can be considerable. Furthermore, despite the use of explicit weights in MCA, the analyst may unintentionally introduce implicit weights during the evaluation process that may lead to results that cannot be explained.

Life-Cycle Analysis (LCA) involves making detailed measurements of input use and environmental waste during the production of a product, from the mining of the raw materials used in its production and distribution through to its use, possible reuse or recycling, and its eventual disposal. LCA has thus far focused on the environmental burden a product poses throughout its life. It offers possibilities for extension to economic performance, when combined with the life-cycle cost-assessment method (Azapagic and Clift 1999; Hagelaar and Van der Vorst 2002; Carlsson-Kanyama et al. 2003). Using the life-cycle cost-assessment method it is possible to integrate economic and environmental cost information into the LCA framework and assess the cost and environmental effects associated with the life cycle of a product or process. The advantage of this method lies in the fact that LCA allows the establishment of comprehensive baselines of information on a product's or processor's resource requirement. Secondly, it allows identifying areas within a product's life cycle where the greatest reduction of environmental burdens can be achieved. LCA has two main disadvantages. First, it is a data-intensive methodology. Second, the proliferation of conflicting life-cycle analyses on the same products (environmental indexes assigned to each type of material can be influenced by the criteria and priority in developing the indices) are causing customers' confusion and a lack of confidence in the LCA methodology.

Hagelaar and van der Vorst (2002) used Life Cycle Assessment (LCA) to structure environmental supply chains. Their main objectives were: 1) to develop guidelines for managers of supply chains from an environmental perspective; 2) to relate a supply chain to its environmental performance; and 3) to assess the applicability of LCA as a tool for environmental supply-chain management. They concluded that if chains use LCA as a management instrument, they may have to adjust the chain structure to meet requirements set for the use of that instrument. In their paper they argue that in line with a differentiation between environmental-care chain strategies and environmental chain performances, a differentiation between types of LCA should be made, i.e., between compliance-process and market-oriented LCAs. To execute these different types of LCAs, the chain structure should be adjusted to meet the specific requirements of these types. They found that the choice of the type of LCA is conditional on factors external and internal to the chain such as competition, governmental laws, consumer preferences (external) and budget, knowledge, technology, cooperation (internal), etc. Thus the integration of different types of LCAs in the chain brings about a different chain structure.

Data Envelopment Analysis (DEA) measures the efficiency of a firm (chain) relative to the efficiency of competitors. The problem with respect to efficiency in supply chains is that beside direct outputs, which are delivered directly to the market, a firm also produces output that is input to a firm in the next stage. These intermediate outputs are intermediate inputs to the firm in the adjacent stage, next to the direct inputs. Contributions of Zhu (2003) in this field are a first step to wards measuring supply-chain efficiency. The method allows inclusion of various

Table 2. *Advantages and disadvantages of methods to assess supply-chain performance*

Methods	Advantages	Disadvantages
Activity-Based Costing (ABC)	• Gives more than just financial information • Recognizes the changing cost behaviour of different activities	• Costly data collection • Difficulties to collect initially required data • Difficulties to determine appropriate and acceptable costs drivers
Balanced Scorecard	• Balanced view about the performance • Financial and non-financial factors • Top-level strategy and middle-management-level actions are clearly connected and appropriately focused	• Not a quick fix • Complete implementation should be staged
Economic Value-Added (EVA)	• Considers the cost of capital • Allows projects to be viewed separately	• Computation difficulties • Difficult to allocate EVA among divisions
Multi-Criteria Analysis (MCA)	• A participatory approach to decision-making • Enables decision-maker to learn more about the problem • Suitable for problems where monetary values of the effects are not readily available	• Information requirements to derive the weights can be considerable • Possibility to introduce implicit weights leading to results that cannot be explained
Life-Cycle Analysis (LCA)	• Allows to establish comprehensive baselines of information on a product's or processor's resource requirement • Allows to identify areas where the greatest reduction of environmental burdens can be achieved • Possibility to assess the cost and environmental effects associated with the life cycle of a product or process	• Data-intensive methodology • Lack of confidence in the LCA methodology
Data-Envelopment Analysis (DEA)	• All inputs and outputs are included • Generates detailed information about the efficient firms within a sample • Does not require a parametric specification of a functional form	• Deterministic approach • Data-intensive
Supply-Chain Council's SCOR® Model	• Takes into account the performance of the overall supply chain • Balanced approach • Performance of the supply chain in multiple dimensions	• Does not attempt to describe every business process or activity • Does not explicitly address training, quality, information technology and administration

dimensions, e.g., economic and environmental performance. The problem with measuring supply-chain efficiency using the DEA model is that it requires an enormous amount of data, while data gathering is one of the most complex issues in a supply-chain context. The advantages of DEA modelling are numerous. DEA takes a systems approach, which means that it takes into account the relationship between all inputs and outputs simultaneously. DEA generates detailed information about the efficient supply chain within a sample and which supply chains can be used as a benchmark. DEA does not require a parametric specification of a functional form to construct the frontier. Thus there is no need to impose unnecessary restrictions on the functional forms that very often become a cause of distorted efficiency measures. DEA has the disadvantage of being a deterministic approach, which implies that statistical noise may be confounded with inefficiency.

Talluri et al. (1999) studied the importance of a partner selection process in designing efficient value chains. They propose a two-stage framework, where the first stage involves identification of efficient candidates for each type of business process (manufacturing, distribution, etc.) using DEA and the second stage encompasses the use of an integer goal-programming model to select an effective combination of the efficient business processes. Talluri and Baker (2002) proposed a multi-phase mathematical programming approach for effective supply-chain design. They developed a combination of multi-criteria efficiency models based on game-theory concepts and linear integer-programming methods. The first phase evaluates suppliers, manufacturers and distributors in terms of their efficiencies with respect to input used and output generated. The model developed in this phase is a combination of a DEA model and a Pair-wise Efficiency Game (PEG). These methods generate an efficiency score for each candidate. The second phase includes the application of an integer-programming model, which optimally selects candidates for supply-chain network design by integrating efficiency scores from the first phase, demand and capacity requirements, and location constraint. The third phase identifies the optimal routing for all individuals in the network by solving a minimum-cost transhipment model.

It is clear from Table 2 that all described methods have their advantages and disadvantages. Therefore, there is a need to consider carefully all arguments for and against the selected method to measure supply-chain performance. It is also possible to combine two different methods to measure supply-chain performance. For instance, Balanced Scorecard can be combined with EVA, because the EVA method is project-focused, while Balanced Scorecard is functional-focused. Nevertheless, when using a combination of different performance measurement methods, great care needs to be taken to avoid conflicts between different performance matrices used to evaluate the performance of the chain in different dimensions.

AGRI-FOOD SUPPLY CHAIN

When developing a supply-chain measurement system it is imperative to consider the supply chain to be measured since it may have specific characteristics. In general two types of agri-food supply chains can be distinguished: 1) supply chains for fresh products such as fresh vegetables, flowers and fruit; 2) supply chains for processed food products such as canned food products, dessert products, etc. This research is focused on supply chains for fresh agricultural products, more specifically on vegetable supply chains. These supply chains consist of growers, auctions, wholesalers, importers and exporters, and retailers. The main processes are producing, storing, packing, transportation and trading of these products. These supply chains have many specifications, which set them apart from other types of supply chains. Several authors (Van der Vorst 2000; Van der Spiegel 2004) have summarized the following specific aspects of agri-food supply chains:

1. shelf-life constraints for raw materials and perishability of products, intermediates and finished products, and changes in product quality level while progressing through the supply chain (decay);
2. long production throughput time (production of new or additional products requires a long time);
3. seasonality in production;
4. seasonal supply of products requires global sourcing;
5. conditioned transportation and storage required;
6. variable process yield in quantity and quality due to biological variations, seasonality, factors connected with weather, pests and other biological hazards;
7. storage-buffer capacity restrictions, when materials or products can only be kept in special containers;
8. governmental rules concerning environmental and consumer-related issues (CO_2 emission, food-safety issues);
9. physical product features like sensory properties such as taste, odour, appearance, colour, size and image;
10. additional features: e.g., convenience of ready-to-eat meal;
11. product safety: increased consumer attention concerning both product and method of production: no risks for the consumer of foods are allowed;
12. perceived quality, also relevant for food applications: e.g., advertisement or brands (marketing) can have a considerable influence on quality perception.

Recent socioeconomic developments have resulted in a change in performance requirements for food supply chains as a whole and for all stages in the supply chain (Van der Vorst 2000). This change is the outcome of the variation in buying behaviour of consumers. Consumer preferences have become the major determinant of quality and production methods. Food safety and human health are important social concerns, particularly when it comes to greenhouse vegetables (Buurma 2001). Consequently, demand for fresher products and products with higher added values increases. The use of pesticides and other chemicals negatively affects consumers' buying behaviour. Consequently, consumers have high demands on a broad range of quality aspects like food safety, production characteristics, sensory properties, shelf life, reliability, convenience, availability and quality/price ratio (Van der Spiegel 2004). The risks associated with poor quality (e.g., outbreaks of animal diseases and low food safety) are so high that retailers and consumers claim to be increasingly prepared to pay more for higher quality (Van der Vorst et al. 2001). Nonetheless, 'price wars in supermarkets that are vying for consumers' loyalty and international competition are putting pressure on prices. Furthermore, regardless of all the demands for specific attributes, many consumers around the world remain price buyers.

Agri-food supply chains are very sensitive to policy changes concerning the environmental issues. During the past 7-10 years, in The Netherlands public concerns arose about the production system for greenhouse vegetables (Buurma 2001). These concerns were associated with pollution, industrial processes and bulk production. The government took responsibility and covenants were concluded to reduce the use of pesticides and energy by 50%. Besides the consumers' preference variation, environment plays a crucial role in agri-food supply-chain performance assessment, because agricultural products are strongly influenced by nature. The environmental variability (e.g., weather conditions) can be reflected in the quantity and the quality of the farm products. The perishability of fresh products such as fruits and vegetables put strains on logistics and quality management. Given these facts we can say that food quality and environmental issues have a great impact on agri-food supply-chain performance. Thus, based on the specifications of agri-food production, when developing a performance measurement system for agri-food supply chains, the indicators that reflect the quality aspects of products and processes are highly relevant (freshness, food safety, environmental issues, etc.) and together with other financial and non-financial indicators, included into one performance measurement system.

Quality is difficult to define and therefore difficult to measure. The quality indicators of a product in literature are often divided into intrinsic and extrinsic quality attributes (Jongen 2000; Luning et al. 2002; Tijskens 2004) or similarly into product and process quality indicators (Northen 2000). For years, performance of production systems has commonly been evaluated by measuring costs or by measuring the intrinsic product quality such as product safety and sensory properties (taste, colour, texture) (Van der Spiegel 2004). Quality is a multidimensional construct that is based on both perceived intrinsic and extrinsic quality attributes available in the shop (Acebron and Dopico 2000). This means that a buying decision is based on more than only intrinsic properties of a product; extrinsic properties also play a role.

Intrinsic quality indicators refer to physical properties such as flavour, texture, appearance, shelf life and nutritional value. The properties are directly measurable and objective. Quality is formed by turning physical properties of a product into quality attributes by the perception of the consumer (Jongen 2000). The intrinsic product properties define the state of the product, which is evaluated with respect to quality criteria imposed by a producer or user (Sloof et al. 1996).

Extrinsic quality attributes refer to the production system and include factors such as the amount of pesticides used, type of packaging material, use of biotechnology (Jongen 2000). Extrinsic factors do not necessarily have a direct influence on physical properties but influence the acceptance of the product for consumers. The total of intrinsic and extrinsic factors determines the purchase behaviour (Jongen 2000).

In this study we follow the division according to the division into intrinsic (product) and extrinsic (process) quality indicators by Luning et al. (2002). In their work, Luning et al. have divided product quality into 3 aspects: 1) food safety and health; 2) sensory properties and shelf life; 3) product reliability and convenience. Process quality also consists of 3 aspects: 1) production system characteristics; 2) environmental aspects; 3) marketing. Within product safety and health, health refers to food composition and diet. Food safety refers to the requirement that products must be 'free' of hazards with an acceptable risk. The sensory perception of food is determined by the overall sensation of taste, odour, colour, appearance, texture and sound, which are determined by physical features and chemical composition. The shelf life of a product can be defined as the time between harvesting or processing and packaging of the product and the point at which it becomes unacceptable for consumption. Product reliability refers to the compliance of actual product composition with product description, and convenience relates to the ease of use or consumption of the product for the consumer (Luning et al. 2002). Production system characteristics refer to the way a food product is manufactured and includes factors such as pesticides used, animal welfare and use of genetic engineering. Environmental implications of agri-food products refer mainly to the use of packaging and food waste management. Marketing efforts determine quality attributes, affecting quality expectation. Process specifications include the type of equipment needed and handling conditions required. Jongen (2000) and Northen (2000) name traceability and organic production as examples of process indicators.

DEVELOPING A CONCEPTUAL FRAMEWORK

Based on the literature review on existing performance indicators and taking into account the theoretical frameworks underlying the different methods and models such as SCOR® model and/or Balanced Scorecard, the conceptual framework has been developed. The framework takes into consideration specific characteristics of agri-food supply chains. For this purpose, the agri-food supply-chain performance indicators are grouped in four main categories: efficiency, flexibility, responsiveness and food quality. The categories efficiency, flexibility and responsiveness are chosen based on Table 1. These main categories contain more detailed performance indicators. Based on the framework of food quality developed by Luning et al. (2002), the specifications of agri-food supply chains are grouped under the category 'food quality'. Adding the category 'food quality' to the three other categories derived from the literature review results in a complete conceptual framework for measuring the performance of agri-food supply chains (Figure 1).

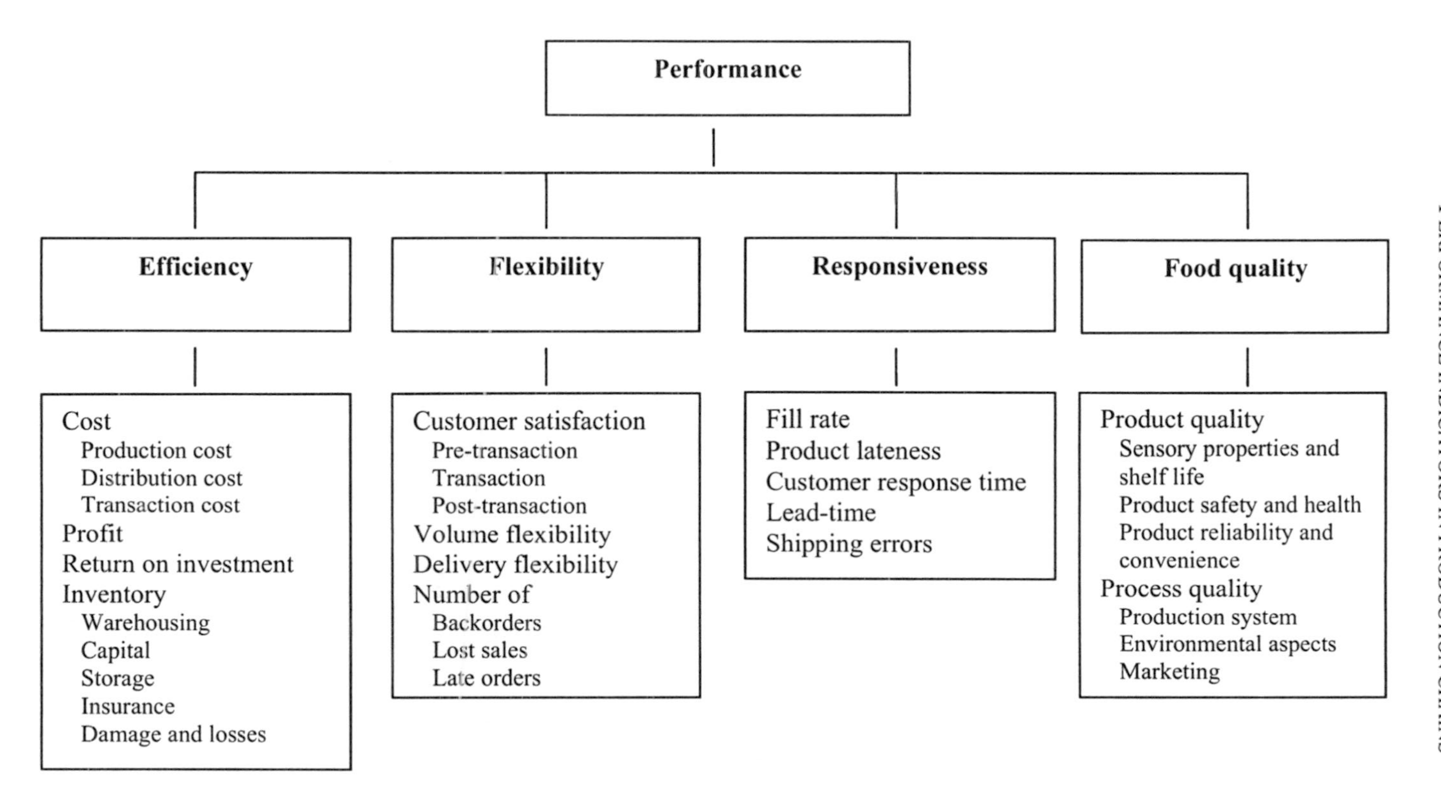

Figure 1. *Conceptual framework of agri-food supply-chain performance indicators*

FUTURE OUTLOOK

This paper reviewed the available supply-chain performance indicators and models and methods used to asses the performance of supply chains. Based on the existing body of research in supply-chain performance measurement systems a research framework has been suggested for measuring the performance of agri-food supply chains. The suggested framework is based on a literature review and needs to be tested empirically. In future research this conceptual framework will be tested by interviewing the experts (managers) and stakeholders across the entire agri-food supply chain. During the interviews experts will be asked to judge the feasibility and the measurability of suggested indicators. Experts will be given the opportunities to suggest new indicators and to reject the proposed ones and to provide suggestions for better (practically possible) ways to measure the suggested indicators. This procedure should be provided with sufficient argumentation. Based on the results of interviews the final research framework for measuring the performance of the agri-food supply chain will be developed that will meet criteria of inclusiveness, universality, measurability and consistency.

NOTES

[1] The work of Hannus (in Finnish) is taken from the paper by Korpela et al. (2002)

REFERENCES

Acebron, L.B. and Dopico, D.C., 2000. The importance of intrinsic and extrinsic cues to expected and experienced quality: an empirical application for beef. *Food Quality and Preference,* 11, 229-238.

Azapagic, A. and Clift, R., 1999. The application of life cycle assessment to process optimisation. *Computers and Chemical Engineering,* 23 (10), 1509-1526.

Beamon, B.M., 1998. Supply chain design and analysis: models and methods. *International Journal of Production Economics,* 55 (3), 281-294.

Beamon, B.M., 1999. Measuring supply chain performance. *International Journal of Operations and Production Management,* 19 (3/4), 275-292.

Berry, D. and Naim, M.M., 1996. Quantifying the relative improvements of redesign strategies in a PC supply chain. *International Journal of Production Economics,* 46, 181-196.

Bijman, W.J.J., 2002. *Essays on agricultural co-operatives:go vernance structure in fruit and vegetable chains.* Proefschrift Rotterdam [http://www.lei.wageningen-ur.nl/publicaties/PDF/2002/PS_xxx/PS_02_02.pdf]

Buurma, J.S., 2001. *Dutch agricultural development and its importance to China; case study: the evolution of Dutch greenhouse horticulture.* Lei, The Hague. [http://www.lei.wageningen-ur.nl/publicaties/PDF/2001/6_xxx/6_01_11.pdf]

Carlsson-Kanyama, A., Ekstrom, M.P. and Shanahan, H., 2003. Food and life cycle energy inputs: consequences of diet and ways to increase efficiency. *Ecological Economics,* 44 (2/3), 293-307.

Claro, D.P., Hagelaar, G. and Omta, O., 2003. The determinants of relational governance and performance: how to manage business relationships? *Industrial Marketing Management,* 32 (8), 703-716.

Coronel, P., 1998. *Balanced scorecard, performance measurement and reporting.* Available: [http://www.benchmarkingplus.com.au/perfmeas.htm] (2005).

Eppen, G.D., 1979. Effects of centralization on expected costs in a multi-location newsboy problem. *Management Science,* 25 (5), 498-501.

Gunasekaran, A., Patel, C. and McGaughey, R.E., 2004. A framework for supply chain performance measurement. *International Journal of Production Economics,* 87 (3), 333-347.

Gunasekaran, A., Patel, C. and Tirtiroglu, E., 2001. Performance measures and metrics in a supply chain environment. *International Journal of Operations and Production Management,* 21 (1/2), 71-87.

Hagelaar, G.J.L.F. and Van der Vorst, J.G.A.J., 2002. Environmental supply chain management: using life cycle assessment to structure supply chains. *International Food and Agribusiness Management Review,* 4, 399-412.

Hobbs, J.E. and Young, L.M., 2000. Closer vertical co-ordination in agri-food supply chains. *Supply Chain Management,* 5 (3), 131-142.

Jongen, W.M.F., 2000. Food supply chains: from productivity toward quality. *In:* Shewfelt, R.L. and Brückner, B. eds. *Fruit & vegetable quality:an integrated view*. Technomic, Lancaster, 3-18.

Kaplan, R.S. and Norton, D.P., 1992. The balanced scorecard: measures that drive performance. *Harvard Business Review,* 70, 71-79.

Korpela, J., Kyläheiko, K., Lehmuswaara, A., et al. 2002. An analytic approach to production capacity allocation and supply chain design. *International Journal of Production Economics,* 78 (2 Special Iss.), 187-195.

Lai, K.H., Ngai, E.W.T. and Cheng, T.C.E., 2002. Measures for evaluating supply chain performance in transport logistics. *Transportation Research. Part E Logistics and Transportation Review,* 38 (6), 439-456.

Lambert, D.M. and Cooper, M.C., 2000. Issues in supply chain management. *Industrial Marketing Management,* 29 (1), 65-83.

Lapide, L., 2000. What about measuring supply chain performance? *ASCET,* 2.

Lee, H.L. and Billington, C., 1992. Supply chain management: pitfalls and opportunities. *Sloan Management Review,* 33, 65-73.

Li, D. and O' Brien, C., 1999. Integrated decision modelling of supply chain efficiency. *International Journal of Production Economics,* 59 (1/3), 147-157.

Luning, P.A., Marcelis, W.J. and Jongen, W.M.F., 2002. *Food quality management:a techno-managerial approach*. Wageningen Pers, Wageningen.

Murphy, G.B., Trailer, J.W. and Hill, R.C., 1996. Measuring performance in entrepreneurship research. *Journal of Business Research,* 36 (1), 15-23.

Northen, J.R., 2000. Quality attributes and quality cues: effective communication in the UK meat supply chain. *British Food Journal,* 102 (3), 230-245.

Persson, F. and Olhager, J., 2002. Performance simulation of supply chain designs. *International Journal of Production Economics,* 77, 231-245.

Romero, C. and Rehman, T., 2003. *Multiple criteria analysis for agricultural decisions*. Elsevier, Amsterdam.

Sloof, M., Tijskens, L.M.M. and Wilkinson, E.C., 1996. Concepts for modelling the quality of perishable products. *Trends in Food Science and Technology,* 7, 165-171.

Stephens, S., 2000. The supply chain council operations reference (SCOR) model: integrating process, performance measurements, technology and best practice. *Logistics Spectrum* (Jul-Sep), 16-18. [http://www.findarticles.com/p/articles/mi_qa3766/is_200007/ai_n8909126]

Supply-Chain Council, 2004. *SCOR*. Available: [http://www.supply-chain.org/index.ww] (2004).

Talluri, S. and Baker, R.C., 2002. A multi-phase mathematical programming approach for effective supply chain design. *European Journal of Operational Research,* 141 (3), 544-558.

Talluri, S., Baker, R.C. and Sarkis, J., 1999. A framework for designing efficient value chain networks. *International Journal of Production Economics,* 62 (1/2), 133-144.

Thonemann, U.W. and Bradley, J.R., 2002. The effect of product variety on supply-chain performance. *European Journal of Operational Research,* 143 (3), 548-569.

Tijskens, P., 2004. *Discovering the future:modelling quality matters*. Proefschrift Wageningen [http://www.library.wur.nl/wda/dissertations/dis3586.pdf]

Van der Spiegel, M., 2004. *Measuring effectiveness of food quality management*. Proefschrift Wageningen

Van der Vorst, J.G.A.J., 2000. *Effective food supply chains:generati ng, modelling and evaluating supply chain scenarios*. Proefschrift Wageningen [http://www.library.wur.nl/wda/dissertations/dis2841.pdf]

Van der Vorst, J.G.A.J., Van Dijk, S.J. and Beulens, A.J.M., 2001. Supply chain design in the food industry. *The International Journal on Logisitics Managment,* 12 (2), 73-85.

Wang, N., 2003. *Measuring transaction costs:an incomplete survey* . Ronald Coase Institute, Chicago. Ronald Coase Institute Working Papers no. 2. [http://coase.org/workingpapers/wp-2.pdf]

Zhu, J., 2003. *Quantitative models for performance evaluation and benchmarking: data envelopment analysis with spreadsheets and DEA Excel solver.* Kluwer, Dordrecht. International Series in Operations Research & Management Science no. 51.

SHARING COSTS, BENEFITS, AND RISK IN AGRI-FOOD CHAINS

CHAPTER 6

RETAILERS' SUPPLY CHAIN, PRODUCT DIFFERENTIATION AND QUALITY STANDARDS

ÉRIC GIRAUD-HÉRAUD AND LOUIS-GEORGES SOLER[#]

[#] INRA-LORIA, 65 Boulevard de Brandebourg, 94205 Ivry-sur-Seine, France
E-mail: soler@ivry.inra.fr

Abstract. The growth of Private Label brands in the sector of fresh agricultural products is a recent occurrence closely related to the food and food-safety crises of recent years. While the public authorities were creating new control and health-monitoring procedures, tightening regulatory production standards and enhancing regulations related to official marks of quality, some retailers were adopting new segmentation strategies for demand. How have these strategies changed the demand for food? To what extent have they altered retailer–producer relationships and under what conditions would it be beneficial for the involved parties to make a commitment? How do these strategies interact with those of the public authorities?
Keywords: food safety; quality; food-processing chains; brands; retailers

INTRODUCTION

The food-processing industry has been faced for several years with increasingly strong consumer demands in relation to product quality and safety. Recent food safety crises, especially the one of the mad-cow disease, have resulted in a loss of consumer confidence. The most notable manifestation of this phenomenon has been a 25 % drop in the consumption of beef in the European Union at the end of the year 2000. These events clearly demonstrated the weakness of the existing mechanisms designed to guarantee food quality and safety. In addition, the public authorities and private operators were prompted to take action in order to (i) define production processes to reduce health risks as much as possible and (ii) set up a system to control and certify the implementation of these processes by the firms throughout the entire production, transformation and commercialization chains. The action taken by the French public authorities was two-fold:

- Firstly, the creation of public agencies entrusted with the task of monitoring public health (the AFSSA – French Food Safety Agency – was founded in 1999 and a European Food Agency in 2001), and the founding of certification and

C.J.M. Ondersteijn et al. (eds.), Quantifying the agri-food supply chain, 69-85.

control agencies responsible for ensuring compliance with standards and product quality specifications.

- Secondly, the defining of minimum quality and food safety standards whose application is mandatory for all relevant parties. The best-known example is certainly the ban on bone meal for cattle feed. Many other initiatives were taken, especially in application of EU Directive 89/397, which requires that national law ensure a regular control over production, the standards recognized by all Member states and the annual submission to the European Commission of national legislation regarding the food sector (Fearne 1998).

Concurrently with these events, private initiatives have been taken in order to stem the sharp drop in consumption. The brand image of major retail firms suffered greatly as a result of the mad-cow crisis, and retailers have sought to respond to consumer expectations, not only in terms of product safety but also sensory and environmental quality[1]. In the United Kingdom a number of different approaches have been adopted by the country's largest retailers; for example, a 'Traditional Beef' approach developed by Sainsbury or 'Select Beef' developed by Marks & Spencer. In France direct agreements have been concluded between retailers and producers under the banner of collective bodies whose purpose is to concentrate demand and guarantee the application of specific product quality specifications. This has been the case with Carrefour, which in recent years has concluded supply contracts with producer organizations under the name of 'Filière qualité Carrefour' (Carrefour Private Certification Chain Brand). More than 60 fresh food products have been included in these agreements, and ultimately this figure should rise to some 200 products. Likewise, Auchan and its policy of 'Responsible Agriculture' serve as an illustration of this approach. Agreements have been made in 30 food chains in recent years and now 100 products have been included. The declared objective is to have 80% of the fruit and vegetable supply and 25% of fresh food products in compliance with this system[2]. Fundamentally, these approaches reflect the desire of the retailers to provide consumers with a guarantee as to the safety and quality of products, in a sector which traditionally has had very little, if any, segmentation, and in which uncertainty about product characteristics on the market has been amplified by recent crises (Sans and De Fontguyon 1999).

An important aspect of these new forms of production and commercialization of food products is that they tend to be based on what shall be referred to as 'chain brands' in this paper. The aim of these chain brands, or private certification labels, is to associate, at least partially, upstream agricultural players with the certification of the products supplied to consumers. Chain-brand products are directly related to the more traditional 'private label' products. However, in the case of private labels, product labelling is solely linked to the retail firm that distributes the product. The aim of this article therefore is to present the new key economic issues related to chain brands.

This article will examine several crucial points relative to this subject: (1) Retailers are implementing these systems to a greater proportion of food supply for consumers wishing to have a higher degree of food quality and safety. What is the benefit of such actions? (2) How has the producer–retailer relationship changed and what are the necessary conditions for the parties involved to adhere to this system? (3) How do these initiatives interact with the measures taken by the public authorities, in particular the raising of Minimum Quality Standards?

In Section 2 we describe the new strategies relative to product supply that have been adopted by the retailers, and the key principles on which they are based. We focus a great part of our attention on a case studied in the beef sector, which was gravely impacted by the recent food safety crises. We show that this relationship is grounded on a more 'cooperative' relationship than in previous food supply-system models, in particular because the retailer accepts to forfeit a part of the flexibility afforded by the use of spot markets. However, in exchange, retailers are able to communicate to consumers about the production conditions of the products sold in their retail outlets. In Section 3 we analyse the contractual risks related to these new strategies and the proposed solutions put forward by the producers and retailers. In Section 4 we examine the underlying foundation of the commitment in this type of contractual relationship from the standpoint of both the retailers and suppliers. This point will especially be discussed in relation to the creation and sharing of value. Finally, in Section 5 we examine the usefulness of these private initiatives designed to strengthen the Minimum Quality Standards defined by the public authorities. Section 6 presents and discusses the conclusions of this paper.

PRIVATE-LABEL BRANDS AND CHAIN BRANDS

Article L. 112-5 of the French Consumption Code of Law defines Private Labels in the following manner: "A product may be considered a Private Label if its characteristics have been defined by the firm or group of firms which organize its retail sale, and which own the brand under which the product is sold". The retailer defines the product quality specifications, indicating both the product characteristics and the production techniques. Private Labels were created some 20 years ago and for quite a long time were used solely for products transformed by the food-processing industry. Private Labels were positioned in price segments equivalent to or below A-brands of those food processors. The creation of Private Labels resulted in a change in the balance of power in the supply chain, in favour of the retail industry.

Private Labels were developed for processed products, but until recently had been absent from the fresh-agricultural-product sectors. In these sectors the system operated as described below:

- Little differentiation existed on the fresh-agricultural-products market, e.g. meat, fruits and vegetables. Private Labels were virtually inexistent and retailers launched products without any particular identifier regarding product origin or quality. Product heterogeneity was at times quite high on the generic product market. However, the performance of this market was such that there was no resulting price differentiation, either at an intermediary level (producer–wholesaler–retailer relationship) or consumer level.
- These products were purchased by the retailers' central purchasing units from a range of intermediaries (e.g. slaughterhouse operators, wholesalers) on specific

spot markets on which the supply and demand relationship was determined on a daily basis. No commitments were made in relation to purchasing or selling between the customers and suppliers.

This total absence of product differentiation and identification suddenly posed a serious problem with the appearance of the mad-cow crisis. The lack of safety guarantees for commercialized products as well as uncertainty about product characteristics prompted consumers to demand greater transparency about the production process. In order to meet consumer expectations in terms of information and commitment by the relevant operators relative to supply and product characteristics, retailers from the outset of the crisis decided to change their purchasing practices and reorganized their supply chains as follows:

- Retailers urged producers to collaborate collectively in the form of associations or collective bodies, and concluded supply contracts with these newly formed organizations. The purpose of these direct agreements is to create 'safe' production groups, thereby providing consumers with products having a higher level of guarantees in terms of quality and food safety.
- Producers apply product quality specifications (imposed by the retailer or defined in common) for the part of the production delivered to the retailer. These product quality specifications may be submitted to the Ministry of Agriculture for official state approval in order to receive certification known as 'Product Conformity Certification'. The granting of this certification denotes an official label of quality and requires the implementation of an outside audit process via an independent certification organization. Upon application the certification enables the retailer to communicate information about production process characteristics to consumers.

For 'sensitive' products, retailers then segment the supply of products present on their shelf space. Supply is frequently composed of two products: (i) one sold at a low price and presented to consumers as a bottom of the range product; this type of product is not subject to any special agreement with the upstream agricultural parties, and (ii) a differentiated product that is positioned at a higher price and subject to long-term agreements with producers committed to the application of product quality specifications designed to provide consumers with a guarantee of food safety[3]. The retailer develops its communication policy on the basis of this product type, and in this way may reassure consumers about the various safety controls performed throughout the supply chain. This communication policy may involve the group of producers or the production area that supplies the products for the brand. It is for this reason that we use the term 'chain brand', or private certification brand, to designate these new types of relationships between producers and retailers.

In order to illustrate this point, one of the first chain brands to be implemented in the fresh-beef sector will be briefly presented. This chain brand was set up by the retailer Carrefour in conjunction with an association of producers called 'Normandy Cattle Quality Chain Brand', and enables the retailer to supply two products: (i) one that is purchased from intermediaries on traditional markets, and (ii) a differentiated product ('Carrefour Chain Brand') that is supplied according to the terms of the agreement with the group of producers.

The principal aim of the product quality specifications is to guarantee (i) compliance with legal standards, (ii) complete traceability, (iii) food safety for the products, and (iv) the organoleptic quality of the beef. In addition to the regulatory aspects, compliance with the product quality specifications certifies that: the meat comes from a specific breed; specific breeding practices and forms are used; animal feed composed of fodder produced on a farm from approved ingredients was used, devoid of growth hormones (antibiotics); and compliance with stringent food safety standards. Moreover, the product quality specifications guarantee a very stringent selection of carcasses following the slaughtering of the animals, and comply with very precise criteria regarding conformation, fattening, age, weight and a minimum meat maturation period, which is higher than for generic products. The product quality specifications are subject to regular controls performed by a third-party certification body. They generate additional costs for both the producers (upgrade of production unit to regulatory standard, additional production costs) and retailers that pay for the certification of the production units.

Apart from these qualitative criteria, the agreement does not explicitly bind the retailer to purchasing specific quantities each year. Therefore, the group of producers itself defines the quantities to be produced, based on an estimation, which itself depends on the number of outlets included in the agreement and the retailer's commitment to allocate a specific percentage of its shelf space to the sale of the products. In addition, the group of producers retains sufficient flexibility in order to ensure that it is always capable of meeting retail orders.

The animals produced by the cattle breeders and 'theoretically' compliant with the product quality specifications, are not always commercialized under the retailer's brand. The quantities purchased and sold by the producers usually account for 60% of animal production on average. The gap between the quantities produced and sold (in compliance with the product quality specifications) is due to two factors: (i) non-compliance of products, which represents approximately 20% of all the animals raised in accordance with the criteria of the product quality specifications; on average approximately 20% do not meet the quality objectives for technical reasons, and (ii) intentional overproduction aimed at absorbing the fluctuations in retailer orders. The volume of overproduction accepted by the producers, and which ends up being commercialized only at a generic market price, depends on the variability of the retailer's orders. Due to uncertainties in demand on the final market (related to fluctuations in consumption or the behaviour of the competitors), the retailer attempts to maintain a degree of flexibility in relation to the product quantities that are ordered, in order to adapt itself as efficiently as possible to the variations in sales without having to bear the consequences of product shortages or costly overstock.

The retailer's commitment in fact is essentially based on a price indexed on reference prices. Producers are paid on the basis of the average weekly regional spot-market price in addition to some bonus payments related to compliance with the product quality specifications. The surplus is calculated on the basis of the number of animals delivered by the association, and then evenly distributed between all the producers. Nevertheless, penalties may be imposed in accordance with the characteristics of each animal sold.

The resulting price is based on fluctuations in the wholesale-market reference prices. This allows retailers both to avoid the risk of producers withdrawing from the agreement if wholesale prices increase, and to remain in alignment with the supply costs of their competitors if prices decrease. The final price of the products specified in the agreement and displayed on the retailer's shelf space is approximately 10% higher than a generic product.

CONTRACTUAL RISKS

The contractual mechanism described in Section 2 seeks to reconcile a certain degree of commitment, and therefore continuity, in the producer–retailer relationship, while also maintaining a degree of flexibility in order to deal with variations in demand. Nevertheless, this entails some risks for the signatories, which are essentially related to a loss of flexibility resulting from the fact that purchases are no longer made exclusively on the spot market.

A supply contract may prove to be attractive in terms of the creation and sharing of value for the two contracting parties, but may not be possible to implement due to the potential contractual risks. The contracting parties may be forced to renegotiate the initial contract following the appearance of unforeseen factors. In the absence of a third party capable of verifying the effective application of the initial contract (and imposing its application, if necessary), one of the two players may find it beneficial to renegotiate the contract and capture a larger share of the created value, if its power of negotiation increased following the appearance of unforeseen factors. This hold-up mechanism was initially studied by Oliver E. Williamson (Williamson 1975) and results in an unnecessary expense related to specific investments on which no returns are received ex post. Oliver Hart and John Moore (Hart and Moore 1988) formally demonstrated that both a buyer and seller are led to under-invest due to the risk of a hold-up in an 'efficient' situation of an integrated food chain. A paper by Gaucher et al. (2002) provides a concrete application of this contractual issue for food-processing food chains. In a study focusing on the wine-producing industry, these authors showed that one of the possible consequences was a drop in product quality, which could be harmful to consumer interests. This difficulty becomes greater in the following cases:

- the fraction of the investment that may be recovered or redeployed is low;
- the investments made by the supplier have an impact on the valorization of the client, or the investments made by the client have an impact on the supplier's production costs, i.e. the problem of externalities;
- the balance of power may be altered following the appearance of these unforeseen factors, as it will be to the benefit of one of the players to propose a new basis for an agreement, while the other player has lost all its negotiating power.

Within the scope of the producer–retailer relationships studied in the present paper, it is important to examine several points: (i) the possible existence of exclusive contracts; (ii) the type of 'property' related to the product quality specifications, i.e. whether the product quality specifications were registered only by the retailer or jointly held by the retailer and producers; and (iii) the degree of specificity of the investments related to the agreement. The risk of a hold-up exists for both the producers and the retailers.

Another risk of this type is known as 'image capture'. This refers to a producer–retailer agreement concluded within the scope of an official quality label owned by the producers (e.g. an AOC label) and based on the product quality specifications registered by the retailer (possibly more stringent than the AOC requirements). This could enable the retailer to communicate about the product and in doing so only associate the retailer's name and image with the quality label. The producers can therefore find themselves to be 'expropriated' and lose the value associated with the quality label that they themselves initially created. Likewise, a retailer that makes a major investment in promotional campaigns to communicate about specific product quality specifications, may also face a hold-up risk, if after having invested the producers choose to channel their production to another client.

One possible solution put forward in the theoretical literature to resolve hold-up-related issues is the sequentiality of investments. De Fraja (1999) studied the following contract in which the client made its investments prior to the conclusion of the contract. The supplier is therefore in a position to observe these investments, and realizing that the risk of a hold-up for itself has diminished, also makes the necessary investments for the purposes of the contract. In order for the client to be able to make the investments prior to the conclusion of the contract and to avoid being exposed to the risk of a hold-up, the investments must not be very specific to the relationship with the supplier. The working relationships implemented by the retailers clearly reflect this contractual method. Indeed, retailers invest in the promotion of their firm through general communication policies that refer very little to specific products, except for the promotion of a few showcase products designed to lend credibility to safety and quality requirements in the eyes of the consumer. This means that as in the case examined above, marketing investments are made before contracting and are generally not specific to agreements with particular groups of producers. Having taken notice of these commercial investments, the producers accept to commit themselves and to make their own investments since they anticipate future growth, or at least the preservation of business opportunities created by these major marketing investments.

In fact, producers are often dispersed and do not have equivalent resources for communication. For this reason they depend on the retailers' communication policies to preserve their business activity. In order to avoid too great a dependency, a key issue for these producers is to develop concurrently quality labels capable of showing consumers their own efforts in this area. This is one reason that explains the very rapid growth of common producer brands based on the origin of the products (e.g. AOC wines) or on more stringent product quality specifications (quality labels). Producers then accept to make a commitment to this type of relationship with retailers, since in this way they attempt to receive a commercial guarantee that will publicize and increase awareness of the product in question. This is done even if it means keeping prices at the same level for a certain length of time, and even though these prices may soon seem low given the required production efforts. The purpose of this type of commitment is to be in a position – at a later stage – to renegotiate the agreements when the quality has been recognized by consumers.

From the examination of these points it is apparent that what is fundamentally at stake in the present case of food-processing chains is the capacity to inform consumers about efforts made at each stage of the chain in order to guarantee product safety and quality. One of the main points of contention is related to product labelling. In some cases, only a reference to the retailer is made, while in other cases the relationship is more 'cooperative' and the product label includes both the private label and a reference to the group of producers. The greater the vertically cooperative nature of the supply relationship between producers and retailers (until attaining a situation in which both stakeholders are listed on the label), the more the retailer's flexibility diminishes. In situations where products are above all 'trusted products' (i.e. it is difficult for consumers to judge certain product characteristics by themselves, such as the long-term effects on health), the loss of flexibility related to these new types of supply is the compensation (for the retailers) for having the opportunity to develop communication policies that stress production conditions, a task normally associated with the supplier and not the retailer.

CREATION AND SHARING OF VALUE

The economic literature dealing with private label products has grown significantly as over the last 20 years private label brands have come to establish themselves firmly on the market in most developed countries (e.g. Hoch 1996). The aim of these studies is generally to examine to what extent this has contributed to value creation, and how the created value is shared by the different stakeholders. For example, Mills (1995) proposes a model of the producer–retailer relationship in which the producer's brand, or 'national brand', and the retailer's brand are placed in competition with one another on the final market. The author shows how private labels increase retail performance by (i) shifting away sales previously made under national brands to private labels supplied at a lower wholesale price, and (ii) increasing profit margins on national brands. Other studies have pursued this work and have focused on the strategic choice of differentiation (Caprice 2000; Bontems et al. 1999) and legal restrictions related to product supply (Allain and Flochel 2001). Nevertheless, it is important to note that in all these studies, private labels have been positioned in a segment in which the final prices and quality levels are lower than that of the national brand.

The focus of the present paper is on chain brands, or private certification brands, which operate according to a different rationale. Having emerged in a sector in which there are no national brands, the chain brands are less dedicated to altering the balance of power with suppliers than providing support to the segmentation of supply for consumers by developing product chains that are positioned at a higher quality and price level than that of the average quality and price of undifferentiated products. Several important points should thus be considered in order to evaluate the impact of these new strategies in terms of value creation and sharing:

- *The degree of differentiation of the product in comparison with the product available on the spot market.* This degree of differentiation is determined by the product quality specifications imposed by the retailer, and which may vary in stringency, and also generate production costs significantly higher than those of the generic product. Additional costs are also generated by the increased number of controls at all stages of the chain, both for chain operators and outside audit and certification bodies.
- *The price of the final product and the consumer reservation price.* These factors depend on the degree of differentiation in relation to the generic product but also on the communication policy, and therefore marketing investments made to increase the awareness of the product and reassure consumers about its characteristics.
- *The alternative means of selling and purchasing for suppliers and retailers.* Depending on the degree of exclusivity of the relationship, each party may or may not have some alternatives to sell or order the differentiated product via other circuits. These possible alternatives have a twofold impact: from the producer's standpoint, the threat of imposing rationing on the retailer may improve its negotiating power and enable it to capture a greater share of the created value; from the retailer's standpoint, the threat of placing several potential suppliers of the different differentiated products in competition against one another, heightens the retailer's negotiating power.
- *The method for defining prices and the volume of quantities exchanged.* In practice, there is a wide diversity of negotiation methods employed between the producers and retailers with respect to these new supply schemes. These agreements are based on cooperative approaches that are relatively strong. In some cases, the prices and quantities are defined by the retailer; in other cases the price is negotiated by both parties and the quantities are imposed by the retailer; and yet in other cases the prices and quantities are negotiated by the producers and the retailer.

In order to quantify the economic impact of these different points, we proposed a model for the agreements concluded between the producers and retailers in a book written by Giraud-Héraud et al. (2002). This model uses the vertical structure in Figure 1 by assuming a higher quality of the chain brand compared to that of the generic product (see Box 1).

Regarding the impact of the producer–retailer relationship on the type of food supply provided to the final consumer, it should be noted that the important strategic decision to be studied is that of the positioning of the product originating from these new supply sources. This strategic decision is based on the following observation: when the quality level increases, the cost of reaching this quality level also increases, thereby resulting in an increase in the final price and in a reduction of the share of shelf space allocated to the differentiated product. The chain is therefore faced with the following alternative: (i) either the qualitative differentiation with respect to the generic product is low and the gap between the production cost and the final price in comparison with the generic product is low, but a major portion of

Box 1. Modelling of the producer–retailer relationship with chain brands

We assume a set R of producers providing a similar product represented by a quality index, denoted k_0. Parameter k_0 represents the minimum quality standard to which all the producers are subject. Each producer has an identical production capacity $\alpha = K/R$ (where K denotes the total supply capacity in the upstream part of the market) and supplies an intermediary market from which N retailers supply themselves. Each retailer j then supplies a market of size M_j ($j = 1,...,N$).

The intermediary market is assumed to be a competitive market in which price ω_0 is formed, thereby equalizing the supply of upstream producers and downstream retailers. This price ω_0 is imposed to each retailer j. Nevertheless, each retailer is free to choose the quantity it desires according to the demand it receives on the final market. On this final market of size M_j, consumers distinguish themselves with a taste parameter θ in terms of the quality offered and it is assumed that parameter θ is distributed over an interval $[0, \bar{\theta}]$, thus making it possible to take into account the heterogeneousness of consumer tastes. A consumer surplus θ purchasing at price p a unit of a product of quality k is expressed as $S(\theta) = \theta k - p$, thereby defining the difference between the willingness to pay and the actual price paid. This model can thus estimate the behaviour of the players in a situation in which only the generic product is offered to the consumers (benchmark situation) We then study the different possibilities for the implementation of a partnership between the group of producers G and one of the retailers in order to shift away a part of the exchanges made on the intermediary market and create a chain brand, or private certification brand. This chain brand corresponds to a partnership between the group of producers G and the retailer N in order to:

i) offer a product of higher quality k_1 to the consumers ($k_1 > k_0$);

ii) define the quantities x_N and y_N commercialized by the retailer with qualities k_0 and k_1.

In this manner, a group of producers G directly supplies retailer N a part y_N of its production potential, with a higher quality, and the remaining balance $\alpha G - y_N$ is allocated to the spot market. The remuneration price for the upstream producers is calculated on the basis of the Nash solution (assuming that the status quo is the benchmark of the relationship). The model is resolved analytically and is calibrated in accordance with data from surveys conducted in different chains, in particular with data from the 'Market News Department' of the Ministry of Agriculture and Fishing. The data used for the simulations presented in the remaining parts of this paper are related to the beef sector. These data are representative of the different stages of the chain: the price in major retail outlets of the generic product and the differentiated products under retailer chain brands; the price paid to producers on the wholesale markets and within the scope of the chain agreements studied in this paper; cost differentials related to production, control and certification between products, with and without retailer product quality specifications; quantities produced within and outside the scope of the producer–retailer agreement. In this type of analysis, quality levels k_0 and k_1 are directly associated with consumer propensity levels and different production cost levels.
In this situation, the raising of the minimum quality standard is assumed to result in an increase in production costs. It is the impact of this increase in costs on the positioning of the chain brand and on the gains of the producers and retailer that is subsequently evaluated (see Figures 1-4).

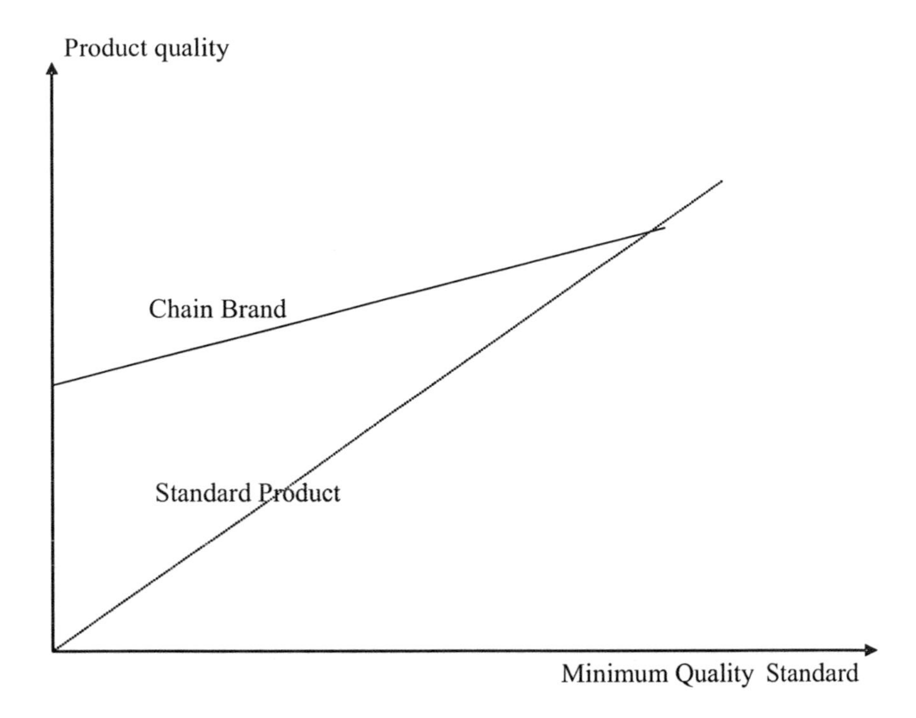

Figure 1. *Shelf space allocation and chain brand quality according to the standard product quality*

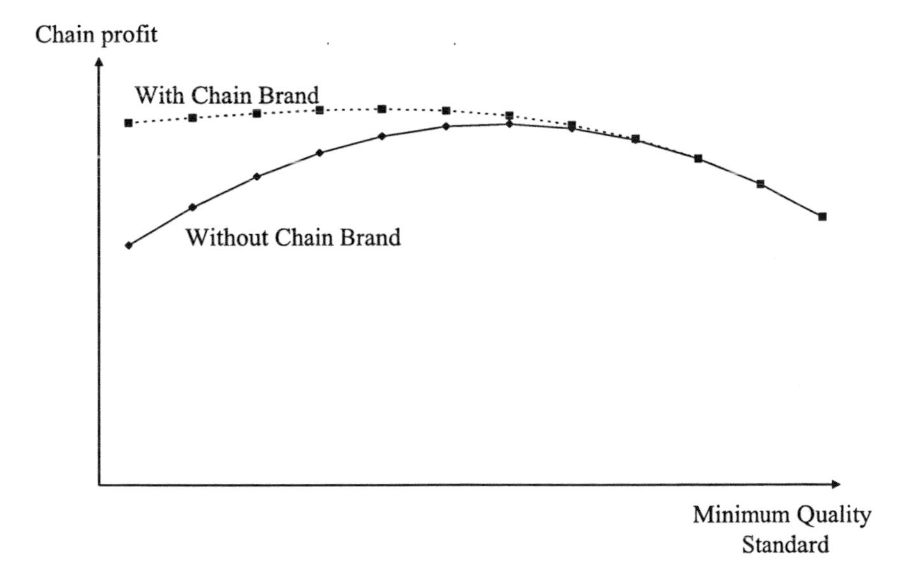

Figure 2. *Chain profit (producers + retailer) according to the Minimum Quality Standard*

the food supply has a slight added value in terms of qualitative and safety guarantees, or (ii) the qualitative differentiation is high, thereby – on the contrary – generating significant additional costs and a difference in final price that is high. In the latter case, the product is perceived as providing a strong guarantee in terms of quality and food safety, and is allocated a small share of the retailer's shelf space. Simulations were conducted for the type of situation which was described in Section 2. Figures 1 and 2 show the results of these simulations.

- Generally speaking, the implementation of these new supply strategies creates value, especially as the relationship between the producers and the retailer is more 'cooperative'. Nevertheless, the lack of the retailer's commitment to specific quantities hinders the effectiveness of this relationship. Indeed, even though this approach allows the retailer to preserve a degree of flexibility in relation to fluctuations in demand, it generates additional costs for the producers, which in turn reduce the overall gain for the chain.
- From the standpoint of the chain, the strategy that creates the most value is the choice of a private label occupying a large part of the shelf space with a moderate quality and price difference. Indeed, studies conducted at retail outlets support this proposition as chain brands represent from 50% to 70% of sales of fresh agricultural products, with a price difference of approximately 10% in relation to generic products. An examination of the product quality specifications reveals that this difference in price is especially due to more stringent controls rather than significantly different production techniques for the generics.

These supply strategies, which to a certain extent reconcile the interests of the producers and retailers, help to upgrade the supply of food provided to consumers in the fresh agricultural products sector through the allocation of significant shelf space to products sold under chain brands, and supplied on the basis of agreements with groups of producers associated with the different retail firms. A more widespread use of this system would result in a profound change in the organization of food-processing chains and in the forms of competition in this sector. It is true that retailers would in fact effectively manage a large part of the production, and would conclude direct partnerships with individual groups of producers. In such a framework, competition would no longer be based on a competitive relationship between retailers, or retailers and producers, but between production–commercialization 'chains'. A retailer and a range of producers would collaborate within each one of these chains.

Can this system be sustained over time? To answer this question, it should be noted that the optimal quality positioning of chain brands naturally depends on the quality level of the generic products. According to our assumptions, the quality level of chain brands increases as the quality level of the generic products rises, but less rapidly. This means that as the quality level of generic products rises, the gap between them and chain brands will diminish. Beyond a certain threshold of quality of the generic product, the chain brand will disappear. In addition, the quality level of the generic product depends on progressive quality improvement related to technical adjustments and also on the Minimum Quality Standards imposed by the public authorities.

MINIMUM QUALITY STANDARDS AND CHAIN BRANDS

In this paper we have seen that the adoption of chain brands, or private certification brands, leads to the selling (on the final market) of a significant number of more tightly controlled products, subject to product quality specifications that are more demanding than those related to less regulated products sold on the spot market, but with a moderate difference in quality. Is this system, based on the initiative of private operators, more efficient in terms of the public's interests? And to what extent does the involvement of the public authorities, in particular with respect to the definition of Minimum Quality Standards (MQS), influence the system?

Numerous theoretical works have studied the issue of MQS and the usefulness of publicly regulating product quality in order to correct certain market imbalances. Indeed, it remains uncertain whether the introduction of MQS could in fact lead to an increase in the average level of quality. In addition, Besanko et al. (1988) demonstrated that the creation of MQS could result in increased prices and less product choice, thereby penalizing a fraction of consumers. The literature has also examined a range of other issues: the decrease in the number of firms, which could be attributed to the creation of MQS (Motta and Thisse 1993), the effects generated by rising costs according to the level of quality produced (Ronnen 1991; Crampes and Hollander 1995) and the strategies adopted by firms in anticipation of a tightening of quality standards (Ecchia and Lambertini 2001; Lutz et al. 2000). However, the theoretical literature remains divided over the usefulness and potential effects of MQS, and the results obtained to date are insufficient to resolve the issue at hand in this paper. In particular, these studies do not consider either the vertical relationship between the firms and their suppliers, or the relative sharing of negotiating power between upstream and downstream firms. Yet it may be assumed that the nature of the vertical relationship influences the sharing of quality costs between suppliers and the retailer, thereby conditioning the quality position of the downstream firm and its response to the introduction of MQS.

A careful examination of simulations of the evolution of gains realized by each type of stakeholder in relation to the MQS level yields the following result:

- *The benefit of product differentiation through the implementation of a chain brand is even higher for the retailer when the MQS level is low* (see Figure 3). For the retailer the tightening of MQS reduces the potential benefit of the chain brands in comparison with a spot-market supply devoid of any contractual restrictions. As MQS rises, it is increasingly important to position chain brands in high-quality segments. However, at the same time its price increases and its share of the retail shelf space diminishes correspondingly. In the end, MQS may disappear when differentiation costs become too high. In other words, the more the public authorities raise the level of requirements regarding MQS in order to respond to consumer fears, the more it becomes difficult for retailers to implement a differentiation strategy.

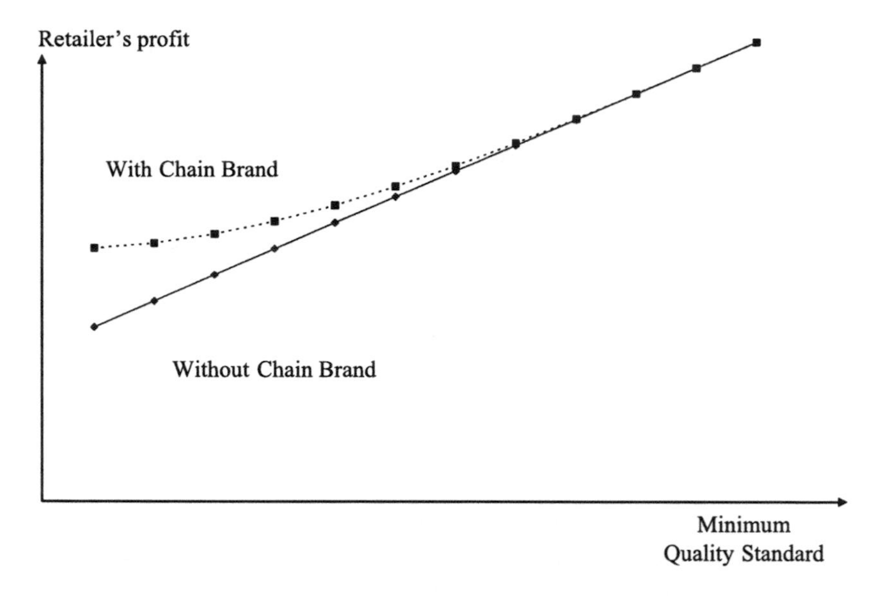

Figure 3. *Retailer's profit according to the Minimum Quality Standard*

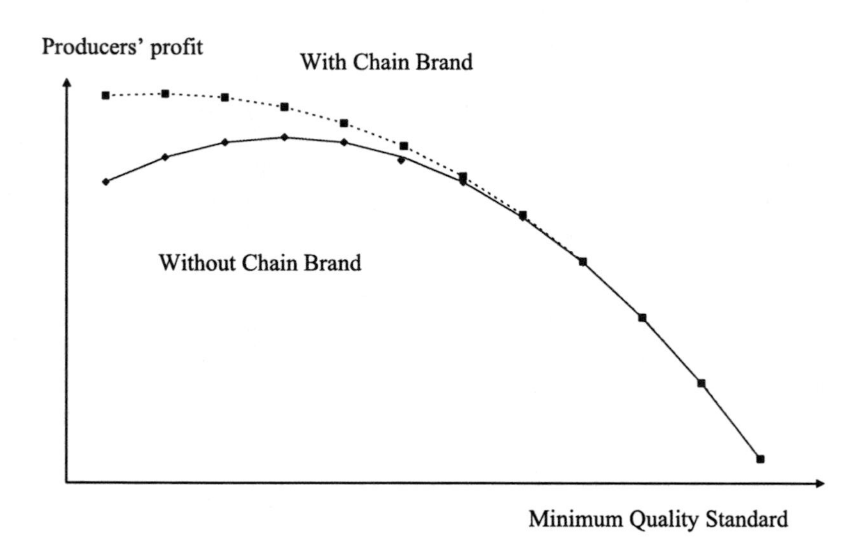

Figure 4. *Producers' profits according to the Minimum Quality Standard*

- *Retailer gains are higher when MQS are high.* In this case, retailers can do without implementing contracts and acquire safe generic products on the spot market. This may be done by placing those suppliers in competition that solely bear the additional production costs of generic products (whereas for chain brands the producer bears the higher costs and the retailer a part of the audit costs).

- *For the producer the enhancement of MQS is favourable to its situation, but generates costs which it must bear and reduces its gains beyond a certain threshold* (see Figure 4). Producers are thus able to maximize their profits when the MQS enable the generic and chain-brand products to coexist on the retail shelf space. In addition, the quality level of generic products that maximize their profits is lower with a chain brand than with a spot market only.

According to the assumptions made in this paper, and regarding consumer behaviour, it should be noted that the surplus (the difference between the propensity to pay given the available quality levels, and the price actually paid, which depends on the available product quantities and qualities) is higher in two differing situations. Firstly, with a low MQS this result in product supply strongly dominated by a chain brand, and with a marked difference in quality with respect to the generic product. Secondly, with a high MQS, this leads to a product supply of generic products throughout the entire retail shelf space (and therefore with no differentiated chain-brand product). However, the consumer surplus is lower than in intermediate situations. The actual balance is closely linked to the costs required for the selected MQS level and those required for chain-brand differentiation.

The issue facing the public authorities when confronted with the mad-cow crisis was to know to what extent they should tighten the norms and standards related to food product quality and safety. Without attempting to define this level of regulation in this paper – as it would require some more advanced and technical arguments – it is interesting to note that one of the key points in the current debate touches upon the costs generated by MQS and the sharing of these costs between the producers, retailers and consumers. Mandatory standards determine the level of production, certification and control costs, and the regaining of consumer confidence presupposes major investments in communication. The balance to be achieved by the public authorities is conditioned by the manner in which the costs are shared and how the profit for each type of player varies. In this paper we have seen that a tightening of MQS, which is thought to discourage the private initiatives of retailers that sell high-quality products, is not necessarily the best solution for these retailers. The development of chain brands is in fact a response to the uncertainties felt by the consumers in terms of the quality level of the generic products. However, a higher quality level of generic products – as a result of more stringent requirements for the MQS – could allow them to withdraw from a long-term relationship with the upstream agricultural players and to return to a system of competitive relationships between producers (even though these stable relations reduce the transaction costs for the implementation of their supply). Paradoxically, we have shown that producers would tend to benefit more from the development of private initiatives by the retailers, and therefore from a moderate increase in the quality level of generics. Indeed, in this situation, a part of the certification and control costs as well as all the communication costs are borne by the retailers. As concerns the consumers, their final decision depends on specific data for the sector and, in particular, the level of production and certification costs imposed by the products in question.

CONCLUSION

The relationship between producers and retailers has led to recurring conflicts for a great number of years, and has prompted the public authorities to adopt new laws on several occasions, in particular to reduce the negative effects of the domination of retailers over producers, which was judged to be too strong. However, as discussed previously in this paper, cooperative strategies are more frequently advocated and put forward within the scope of these new supply systems. It is true that a wide diversity of situations exists and that the degree of 'cooperation' varies according to the sector and retailer. However, communication strategies do stress 'partnerships' with upstream agricultural players. Through the use of these systems the retailers are seeking to reassure consumers by creating their own quality labels and by communicating about the additional guarantees afforded to consumers by these quality labels. These new approaches are causing very profound shifts in the relationship between retailers and the upstream part of the chains. This change reflects a shift away from a relationship between retailers and suppliers solely based on the purchase of products on the spot market (all decisions relating to creation and production are made at the upstream level of the chain) to a situation in which specific agreements are concluded between retailers and producers. These agreements are based on product quality specifications defined by the retail firms and impose quality objectives on which the credibility of their own brands will be judged. One of the most significant impacts of this new type of relationship is that it has created a segmentation of the product supply in a sector in which segmentation previously did not exist at all.

What are the fundamental principles underlying a commitment to such a 'cooperative' relationship, and how are the different operators affected, whether they be in the food-processing sector or they be consumers? Two issues have been examined in order to provide an answer to this question:

- Firstly, the value creation related to this type of relationship, the sharing of this value among the different players of the chain, as well as the quantities and prices of the safe products which are offered to consumers.
- Secondly, the risks related to possible opportunistic behaviours and, in particular, the risk that the efforts required for a substantial enhancement of product quality and safety will not be made at any one of the different stages of the chain.

Should this new type of relationship be considered a long-term phenomenon or should it be viewed as a temporary measure to assuage the fears of consumers, while waiting for all agricultural products to be subject to more stringent controls due to the tightening of MQS? The very rapid development of agreements between producers and retailers, and the considerable investments made by the retail firms to lend credibility in the consumers' eyes to the guarantees provided by this new system would seem to favour a longer-lasting future for these new types of organization. However, the 're-nationalization' of 'responsible agriculture' by the public authorities, and the tightening of the product quality specifications related to quality labels (e.g. AOC in the wine-producing sector) might reverse these commitments.

NOTES

[1] Throughout this article we use the term 'quality' in its economic sense, i.e. the enhancement of the consumer's willingness to pay. In this case, quality may be perceived from both a sensory and an environmental point of view, if the environmentally friendly production conditions favour a higher sales price on the final market.

[2] For more ample details on this topic, we refer to the corporate communication on the retailers' web sites: www.carrefour.fr, www.auchan.com.

[3] Segmentation may comprise up to three segments. For example, in the case of beef, the third segment could be associated with a quality label or an organic product. In addition, there may also be only one segment, in particular for retailers with smaller store layouts, i.e. supermarkets rather than hypermarkets. For more detailed information on this segmentation, we refer to Giraud-Héraud et al. (2002).

REFERENCES

Allain, M.L. and Flochel, L., 2001. Contrainte de capacité et développement des marques de distibuteurs. *Revue Economique,* 52 (3), 643-653.

Besanko, D., Donnenfeld, S. and White, L.J., 1988. The multiproduct firm, quality choice, and regulation. *Journal of Industrial Economics,* 36 (4), 411-429.

Bontems, P., Monier Dilhan, S. and Requillart, V., 1999. Strategic effects of private labels. *European Review of Agricultural Economics,* 26 (2), 147-165.

Caprice, S., 2000. *Contributions àl ânalyse de la puissan ce d'achat dans les relations verticales: interactions stratégiques et marques de distributeur.* Université Paris I, Paris. Thèse de doctorat, Université Paris I

Crampes, C. and Hollander, A., 1995. Duopoly and quality standards. *European Economic Review,* 39 (1), 71-82.

De Fraja, G., 1999. After you sir: hold-up, direct externalities, and sequential investment. *Games and Economic Behavior,* 26 (1), 22-39.

Ecchia, G. and Lambertini, L., 2001. Endogenous timing and quality standards in a vertically differentiated duopoly. *Recherches Economiques de Louvain,* 67 (2), 119-130.

Fearne, A., 1998. The evolution of partnerships in the meat supply chain: insights from the British beef industry. *Supply Chain Management,* 3 (4), 214-231. [http://www.imperial.ac.uk/agriculturalsciences/cfcr/pdfdoc/evolution-of-partnerships.pdf].

Gaucher, S., Soler, L.G. and Tanguy, H., 2002. Incitation à la qualité dans la relation vignoble-négoce. *Cahiers d'Economie et Sociologie Rurales,* 62, 7-40.

Giraud-Héraud, É., Rouached, L. and Soler, L.G., 2002. *Standards de qualité minimum et marques de distributeurs: un modèle d'analyse.* INRA-ESR, Ivry-sur-Seine. Cahier du LORIA no.13.[http://www.inra.fr/esr/UR/ivry/PDF/Cahier13.pdf].

Hart, O. and Moore, J., 1988. Incomplete contracts and renegociation. *Econometria,* 56 (4), 755-785.

Hoch, S., 1996. How should national brands think about private labels? *Sloan Management Review,* 37 (2), 89-102.

Lutz, S., Lyon, T.P. and Maxwell, J.W., 2000. Quality leadership when regulatory standards are forthcoming. *Journal of Industrial Economics,* 48 (3), 331-348.

Mills, D.E., 1995. Why retailers sell private labels. *Journal of Economics and Management Strategy,* 4 (3), 509-528.

Motta, M. and Thisse, J.F., 1993. *Minimum quality standards as an environmental policy domestic and international effects.* Fondazione ENI Enrico Mattei, Milano. Mimeo, Fondazione ENI Enrico Mattei no. 76.95.

Ronnen, U., 1991. Minimum quality standards, fixed costs, and competition. *RAND Journal of Economics,* 22 (4), 490-504.

Sans, P. and De Fontguyon, G., 1999. Choc exogène et évolution des formes organisationnelles hybrides: les effets de la « crise de la vache folle sur la filière viande bovine. *Sciences de la Société,* 46, 173-190.

Williamson, O.E., 1975. *Markets and hierarchies: analysis and antitrust implications:a study in the economics of internal organization.* Free Press, New York.

CHAPTER 7

LIABILITY AND TRACEABILITY IN AGRI-FOOD SUPPLY CHAINS

JILL E. HOBBS

Associate Professor, Department of Agricultural Economics, University of Saskatchewan, 51 Campus Drive, Saskatoon, SK, S7N 5A8, Canada

Abstract. Improving food safety, reducing the impacts of food safety problems, and providing a means to verify food quality attributes are driving the development of traceability initiatives in agri-food systems. Numerous and varied examples exist, from regulatory traceability initiatives, to industry-wide livestock traceability programmes, to individual supply-chain systems that combine traceability with quality verification. This paper explores the economic functions of traceability, examining the extent to which traceability can bolster liability incentives for firms to practice due diligence. The extent to which consumers value traceability per se, versus verifiable quality assurances delivered through traceability, is evaluated empirically using survey and experimental auction data.
Keywords: traceability; food safety; quality verification; credence attribute; experimental auction

INTRODUCTION

Food safety and methods to verify food quality are critical components of modern differentiated food systems. Food safety has garnered significant public policy interest in the wake of highly publicized breakdowns in food safety, particularly those resulting in fatalities (see, for example MacDonald and Crutchfield 1997; Hobbs et al. 2002). For agri-food firms, the implications of a major food safety failure can be commercially devastating, and include: product recalls, damage to reputation and punitive liability damages. Ensuring that acceptable food safety practices are adhered to may require knowledge of actions at prior stages of the supply chain, such as verifying the use of permitted chemical pesticides or food ingredients. In the event of a food safety problem, rapid identification of affected products or batches of products can reduce the number of consumers exposed to a potential harmful foodborne illness.

A highly differentiated food market, with consumers exhibiting diverse preferences, also provides opportunities for firms to gain a competitive advantage through verifying the presence of a desirable quality attribute. Often, this requires identity preservation or quality verification throughout the supply chain, particularly

C.J.M. Ondersteijn et al. (eds.), Quantifying the agri-food supply chain, 87-102.

for process attributes derived from on-farm production methods. It is apparent that, for both food safety and food quality, verifying information flows and identifying positive or negative practices along the supply chain are increasingly important. Traceability of food has become integral to food safety and food quality.

This paper explores the role of traceability in agri-food supply chains, beginning with a discussion of the economic functions of traceability, including the relationship between traceability and liability. It examines the extent to which traceability systems can bolster liability incentives for firms to practice due diligence. The paper shows that, while traceability can perform an important role with respect to strengthening food safety incentives, traceability systems also have an important function in quality verification. The extent to which consumers value traceability per se, versus verifiable quality assurances delivered through traceability, is evaluated empirically using data collected through a survey and an experimental auction.

EMERGING TRACEABILITY SYSTEMS

Food safety incidents, the demand for differentiated food products from an increasingly sophisticated and discerning consumer market, and innovations in quality measurement, tracking and information management technologies have pushed traceability to the forefront of supply-chain issues in the agri-food sector. Traceability is a core component of recent regulatory initiatives in the European Union affecting the entire food and feed sector (for example, Article 18 of the General Food Law, EC 178/2002). It is being used as the basis of competitive product differentiation strategies by food firms seeking to assure consumers of the presence of credence attributes related to production or processing methods. It has been introduced on an industry-wide basis by commodity associations to ensure traceability of livestock from the processing plant to the herd of origin. Thus, traceability appears in many guises, performs many functions and is being driven by many different actors depending on the context. This begs the question: why have traceability? What functions does a traceability system perform? To answer these questions, it is useful to distinguish between the various regulatory, industry-wide and supply-chain-based traceability systems emerging in agri-food sectors, particularly in meat and livestock industries.

Within the EU, traceability has been enshrined in a number of regulatory initiatives. For example, the EU beef-labelling regulation (EC1760/2000) required each member state to introduce a national cattle identification and registration system. The regulation also requires that beef products be labelled with a traceability number at the retail level, enabling the product to be traced back through the supply chain in the event of a problem. Article 18 of the EU General Food Law (EC178/2002) addresses traceability directly, with wide-ranging traceability requirements that came into effect in January 2005[1]. The article requires that traceability of food, feed, food-producing animals and any substance incorporated into food or feed be established throughout the supply chain (from production, through processing and distribution). The regulation requires both upstream and downstream traceability through each adjacent stage of the chain. Specifically, it requires that food and feed business operators be able to identify the supplier of a food, feed or food-producing animal and be able to identify the other businesses to which their products have been supplied. Adequate labelling to facilitate traceability is required. The regulation stops short of specifying how traceability should be ensured, leaving Member States to introduce domestic measures to ensure compliance within their jurisdictions.

Sector-specific traceability initiatives include the introduction of livestock identification programmes that facilitate partial traceback through specific segments of the supply chain. For example, the Canadian Cattle Identification Agency, established in 2001, was largely an industry-driven initiative to put in place a cattle identification system to allow the traceback of cattle from the point of slaughter to the herd of origin. Backed by the Federal Health of Animals Act and Regulations, the system became mandatory in 2002. The Canadian Food Inspection Agency, an agency of the Canadian government, enforces the system and applies penalties where necessary. All cattle must be identified with an approved ear-tag when they move beyond their herd of origin. Similar systems are in place, or being assessed, in other countries.

In 2001, Australia introduced a voluntary National Livestock Identification System (NLIS). Subsequent regulatory initiatives between the state and territory governments, in conjunction with the industry, will move the system to a mandatory status. At its most basic level, the mandatory component of the system will require only that participants identify each animal with an approved tag. However, the traceability infrastructure can also be used to combine animal identification with improved farm management practices and market feedback information through storage of additional information linked to each animal (Meat and Livestock Australia 2005). In the US, the National Cattlemen's Beef Association created an Animal Identification Commission charged with developing a National Animal Identification System (NAIS) for the US beef industry. Concerns over confidentiality, privacy and liability have inhibited the development of a national US cattle traceability system (Beef USA 2005; Souza-Monteiro and Caswell 2004).

Private-sector traceability initiatives are the most sophisticated at the level of the individual supply chain. Individual supply-chain-based traceability initiatives are numerous and varied. These systems have emerged in response to perceived market premiums for quality assurances that can be verified through traceability. An early example was Tracesafe in the UK beef industry (see Fearne 1998), which used a network of cattle breeders and finishers following specific production guidelines, and used traceability as a means of providing a safety and quality assurance to consumers. Other examples include the VanDrie Group in The Netherlands, operating a traceability system for veal, enabling meat cuts to be traced from the retail shelf to the farm of origin and providing information on animal husbandry and production methods (see Buhr 2002).

In Japan, consumer pressure has encouraged a number of supermarkets to implement retail-level traceability capabilities. Through in-store computers or over the Internet, consumers can access information on the source of the beef and the methods used to rear the animal – so called 'story meats' (see Clemens 2003). In Canada, Maple Leaf

Foods, a major processor of fresh and prepared meats and other food products, has identified traceability as an important component of its differentiation strategy for pork products. Early in 2004, the company announced the commercialization of a DNA-based traceability system for its pork products, which is being piloted in the Japanese market (Maple Leaf Foods Inc. 2004). In the event of a future food safety problem in Japan linked to imported pork products, the DNA-based system is also intended to help the company to verify through DNA testing that the affected product did not originate through the Maple Leaf Foods supply chain.

DIVERSE ROLES FOR TRACEABILITY

As the above discussion indicates, traceability systems are emerging in various guises, as a result of both regulatory and industry initiatives. In this context, three key functions of a traceability system can be identified. The first is to allow the efficient traceback of products and inputs (including animals) in the event of a food safety or herd health problem. The primary objective in so doing is to minimize the public and private costs of a problem (Golan et al. 2003; Hobbs 2003; Hobbs et al. 2005). Efficient and timely traceback could limit the size of product recalls or herd quarantine or irradication programmes, and limit the number of people exposed to tainted food, thereby limiting human-health impacts, minimizing productivity losses from illness, etc. The ability to identify and trace affected products or animals may also assist in protecting firms that practice due diligence from free riders. Most national livestock traceability systems, including the Canadian cattle identification system, primarily perform this function. The EU beef-labelling regulation and the traceability article of the General Food Law also primarily perform this function.

Another function of traceability is to reduce information costs for consumers by identifying credence attributes (Hobbs 2003; Hobbs et al. 2005; Golan et al. 2003). For example, this may include labelling of environmentally-friendly production practices, or assurances about feed, other ingredients or production practices. The information requirements tend to be more complex than simple traceability, and are a means through which product differentiation occurs. The private-sector supply-chain-based traceability systems alluded to above, including Tracesafe, the VanDrie system and, to some extent, Maple Leaf Foods, are driven primarily by this motivation. In this respect, traceability is a vehicle through which to deliver quality assurances to consumers that go beyond simple traceback information. Rather than food safety, this function is more broadly linked to verifying quality attributes.

In addition to cost reduction in the event of a food safety problem, a third function of traceability may be as a means of strengthening liability incentives to produce safe food (Hobbs 2003; Golan et al. 2003; Hobbs et al. 2005). If effective, the penalties from statutory or civil liability should discipline firms to practice due diligence with respect to food and feed safety. This potential function of traceability systems is controversial, and may have inhibited the acceptance and adoption of traceability systems among producer groups in some countries, including the USA (Souza-Monteiro and Caswell 2004). Furthermore, it is not clear whether liability could be proven in practice, and therefore whether the threat of liability is an effective incentive (Buzby and Frenzen 1999). Nevertheless, discussions of traceability are often laced with references to liability implications, and it is useful to explore the nature of liability in agri-food systems in more detail.

Liability in the food system

Statutory liability

Liability arises in a number of guises in the food system, and we can distinguish broadly between regulatory or statutory liability and civil or contractual liability. Regulatory liability results from failure to meet mandatory standards and is potentially a criminal offence; for example, if a firm's actions are found to be in violation of food safety legislation that mandates or prohibits specific practices. The penalties to being found liable with respect to a regulatory offence depend on the jurisdiction but generally range from financial penalties to imprisonment. Typically, a government agency monitors compliance with regulatory standards.

In many legal systems, including under Canadian law, for a party to be subject to criminal or statutory liability, two elements of a regulatory offence must be proven. First, it must be proved that the *actus reus*, or guilty action, contained in the offence was committed by the accused. For example, if under the food safety legislation it was an offence to allow food to come into contact with carcinogenic chemicals, and a firm allowed to this occur, the firm/management would have committed the guilty act specified in the offence. Second, it must be proved that there was wilful negligence or recklessness on the part of the accused; this is known as *mens rea* (Wasylyniuk et al. 2003).

In many jurisdictions, regulatory offences may be treated as absolute or strict liability offences for the purposes of prosecution. An absolute liability offence only requires proof that the offence was committed, and allows liability to be imposed without proof of a fault element. However, the strict liability offence is more commonly used, and requires proof that the prohibited act occurred, but bases the fault element on negligence. Once the *actus reus* is proved beyond a reasonable doubt, negligence is presumed, and a reverse onus is placed on the accused party to prove that he or she was not negligent. It is within this context that a due-diligence defence arises, wherein the accused party must show that he or she took all reasonable care to fulfil their legal obligations in meeting the statutory requirement (Wasylyniuk et al. 2003).

The UK 1990 Food Safety Act was notable for extending legal liability to food retailers for the safety of food sold through their stores. Rather than rely on a food manufacturer's warranty in the event of a food safety incident caused by the actions of the manufacturer, retailers are required to show evidence of adequate monitoring of supplies or of suppliers to satisfy their due-diligence defence. This change in regulatory liability had significant implications for supply chain relationships and traceability in the UK food system, encouraging food retailers to form closer and longer-term relationships with their suppliers in order to facilitate monitoring (Hobbs and Kerr 1992).

To return to the hypothetical example of preventing carcinogenic chemicals from coming into contact with food. If the relevant food safety statute in a jurisdiction specified that in order for a party to have committed an offence, the party must have wilfully or recklessly allowed carcinogens to contaminate a food product, it must be proved that this subjective fault (or *mens rea*) was present when the carcinogens were allowed to come into contact with food. However, if the offence does not specify the mental element (wilfulness), the offence could become either absolute liability, wherein the accused would be found guilty regardless of whether they knew about the carcinogenic contamination, or one of strict liability, in which it would have to be shown that negligence played a part in allowing the contamination to occur.

Food safety legislation attempts to deter unfavourable practices, while providing for penalties in the event of an offence under the legislation. Mandatory standards represent an ex-ante set of precautions to limit risk, while monitoring compliance and the application of penalties under the law provide an ex-post method of compensation for harm caused (Wasylyniuk et al. 2003). Avoiding statutory liability requires firms to be aware of, and to comply with, food safety regulations. However, while compliance with a statute absolves a firm of statutory liability, the firm may still be subject to civil liability. Often, the criteria on which civil liability is based are much more general, and the burden of proof is less onerous than is the case with statutory liability.

Civil liability

Firms may be subject to civil liability (often also referred to as contractual or tort liability) for damages for non-criminal acts that injure or cause damage to others. This could include negligence in the production, preparation or handling of food or food ingredients by various parties throughout the supply chain. Liability can also arise in the case of misrepresentation of products. This could include mis-representation of a credence quality attribute to induce a buyer to purchase the product. In practice, it is often difficult to determine fault in civil-liability claims (Boyer and Porrini 2002), particularly in lengthy supply chains where the product passes through a number of stages before reaching the consumer, making it difficult to determine which party was at fault.

Traceability systems that allow food products to be tracked through the food supply chain could assist in determining fault, thereby strengthening the liability incentive for firms to adopt good food safety practices. The effectiveness of liability as a deterrent in the case of food safety practices has been questioned (Buzby and Frenzen 1999; Wasylyniuk et al. 2003). Buzby and Frenzen (1999) argue that there are only limited legal incentives in the US to produce safer food, suggesting that less than 0.01% of cases are litigated, with even fewer paid compensation. The lack of traceability through the US food supply chain may have contributed to this outcome, although Buzby and Frenzen also find that ambiguity about whether microbial contamination is natural or an adulterant has hindered the US legal system from dealing effectively with food safety issues.

Mojduszka (2004) discusses the role that liability insurance can play in encouraging optimal behaviour. In isolation, liability insurance may appear to weaken the incentives to control losses: the primary purpose of ex-post liability is to make the risk imposer pay, whereas the primary role of insurance is to spread risk. However, if insurance premiums are adjusted to reflect the insured firm's behaviour, then insurance encourages efficient decisions (Mojduszka 2004). Insurers have an incentive, ex ante, to screen the firms they intend to insure to guard against adverse selection. They have an incentive, ex post, to monitor the insured to prevent moral hazard.

Whether for statutory or civil liability, it is clear that legal proof of responsibility is essential for liability to be an effective incentive for firms to produce safe, high-quality food. Traceability remains a key element of this proof. However, it is also clear that traceability has a far wider role to play in agri-food supply chains than simply bolstering liability incentives. The product differentiation potential of traceability systems may be the 'carrot' necessary to induce farmers to participate in traceability systems despite liability fears. The introduction of traceability systems, whether regulatory or through the private sector, is often accompanied with rhetoric about consumers demanding more traceability. The remainder of this paper presents empirical results from a study evaluating consumer attitudes toward traceability, food safety and quality assurances.

CONSUMER ATTITUDES TOWARD TRACEABILITY AND PROCESS VERIFICATION

The extent to which consumers value traceability per se, relative to quality verifications about production and process methods, is central to understanding the incentives for firms to structure their supply-chain relationships so as to provide these assurances. While examples are emerging of food products with a traceability assurance, or various quality assurances related to food safety or on-farm production methods, it is rare to find an example of a product that encompasses all three assurances. Furthermore, even in the presence of these products, market data is difficult to access and usually cannot be linked back to individual consumer characteristics. Therefore, an experimental auction and survey were used to gather data on Canadian consumer preferences for traceability, food safety and quality assurance attributes in meat products. Experimental auctions have become a popular tool for obtaining non-hypothetical bids for credence attributes (see, for example, Fox et al. 1994; Hayes et al. 1995; Dickinson and Bailey 2002).

An experimental auction was used to elicit willingness-to-pay bids for beef products with additional assurances regarding food safety, on-farm production methods related to humane animal treatment, and traceability to the farm of origin. Following the experimental auction, participants completed a brief questionnaire gathering socio-economic data and additional stated-preference information. Consumer panels were recruited in two locations, in western Canada (Saskatchewan) and central Canada (Ontario), in 2002. Recruitment was a consumer research company in Ontario, and from a range of demographic groups on the campus of the

University of Saskatchewan. Just over one hundred (104) individuals participated in the beef consumer panels, in groups of 12-14 people[2].

Stated preferences for traceability and quality assurances

The post-experiment questionnaire enabled a direct evaluation of respondents' stated attitudes toward food safety, traceability and process verifications. Participants were asked how much confidence they placed in the Canadian government's current food inspection and safety programme and whether they valued having additional assurances about meat safety, beyond what was currently provided by the Canadian government. Figure 1 indicates that there was a reasonably high degree of confidence among Canadians regarding the food safety regulatory system in Canada. Yet, despite this apparent high level of confidence, Figure 1 also shows that many people indicated that they would still value extra food safety assurances. Two explanations are possible. One, that although the participants are generally happy with the current food safety system, there may be other meat safety assurances that food firms could bolt onto the existing system to differentiate their products further. Alternatively, the results may indicate inconsistency in responses, an inherent weakness in stated-preference surveys. Fortunately, the experimental auction analysis (discussed below) allows us to investigate the robustness of participants' stated preferences in terms of whether they acted on these stated preferences in their revealed preference bidding behaviour.

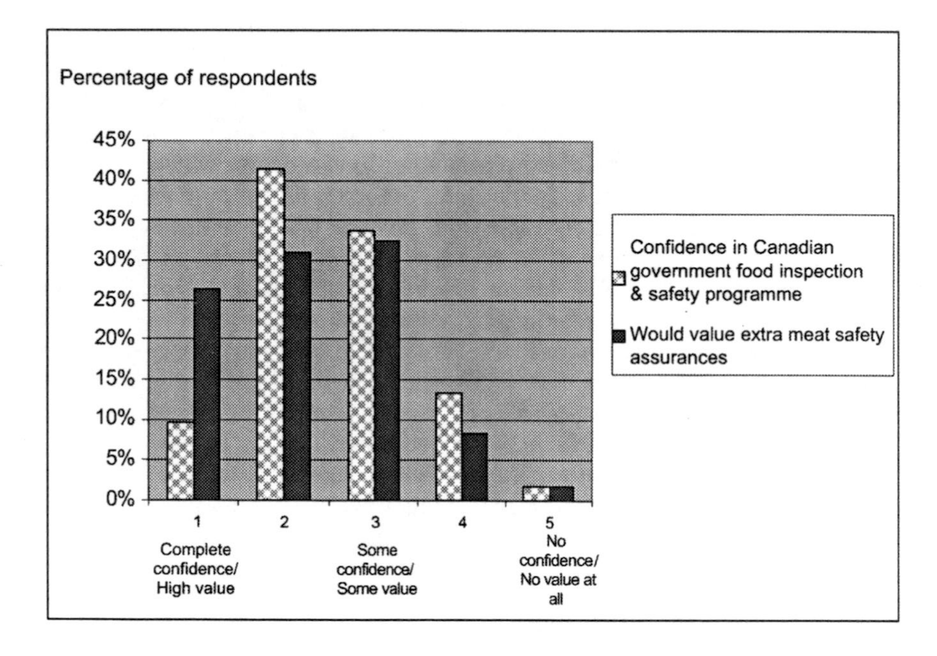

Figure 1. *Confidence in food safety system vs. value of additional assurances (beef) (N=104)*

In the post-experiment survey, participants were also asked whether they would value knowing the exact farm that produced the animals for the meat that they consume. Figure 2 displays the results for the stated preferences with respect to traceability. Only just under one-third (30%) of beef-experiment respondents indicated that this was highly or reasonably highly valued (a score of 1 or 2 out of 5). Again, the experimental auction data allow us to verify whether respondents' stated preferences are supported by their revealed preference bid data.

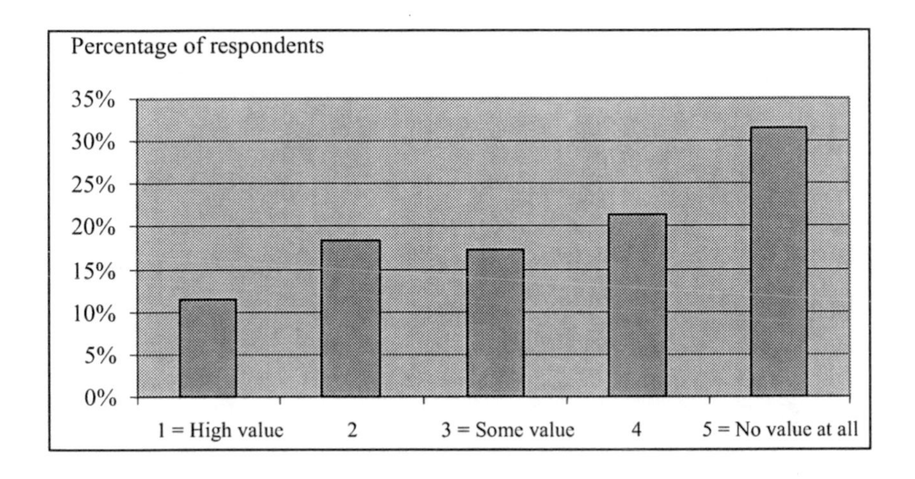

Figure 2. *Value of traceability to the farm of origin (N=104)*

Finally, the post-experiment survey asked whether the participants would value knowing the procedures and processes used by the farmer to produce the animal, such as treatment of the animals, feeds and medication used, presence or absence of genetically modified organisms, etc. It is important to note that traceability systems per se do not necessarily provide additional process verification information. While they may be a vehicle to deliver verification on production methods, simply knowing that the farm of origin could be identified (ex post) does not inform consumers (ex ante) about the production methods used on that farm. Likewise, process verification does not necessarily imply traceability. A downstream food manufacturer may provide an assurance that it only sources products from farms which follow designated production protocols, and may source from a variety of farms, without necessarily being able to trace products back to an individual farm of origin. Therefore, it was of interest to separate attitudes toward process verification from traceability. Figure 3 summarizes the responses to this question.

Relative to simple traceability assurances, receiving information about on-farm production methods appears to be more valuable, with almost 60% of respondents indicating that this type of information would be highly, or relatively highly, valued (scoring 1 or 2 out of 5). The experimental auction data provide a means to verify this finding in the context of actual bidding behaviour.

Respondents who indicated that they would value production method assurances were further prompted to explain why. Almost 60% indicated that they would value this information because it would give them more confidence in the safety and/or quality of the meat they purchased. Twenty-three percent indicated that being able to identify the source of a problem, should one arise, was the primary reason they valued this information.

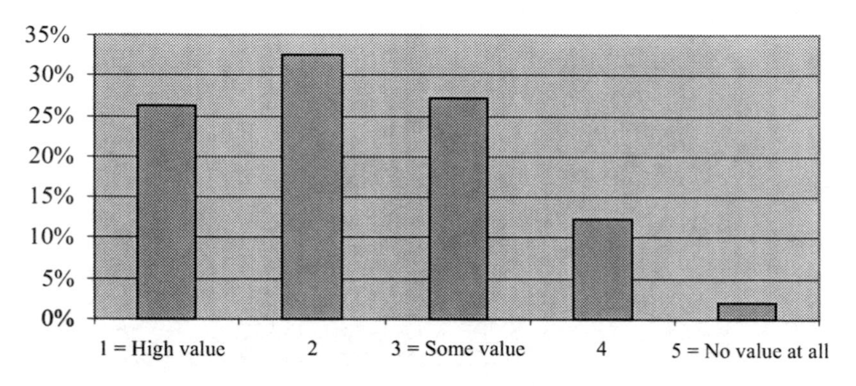

Figure 3. *Value of production method assurances (N=104)*

Preferences revealed through the experimental auction

Prior to answering the survey questions, respondents had participated in an experimental auction to provide a more robust evaluation of the value they placed on traceability and quality assurances. The experimental auction results, briefly summarized here, provide a means to verify the initial findings regarding the relative value of traceability versus other quality assurances[3].

Groups of participants were provided with a beef sandwich as part of a light lunch, and had the opportunity to bid to exchange their sandwich for a sandwich with additional verifiable characteristics, but otherwise identical to their lunch sandwich. Four alternative sandwiches were presented: (i) a sandwich with an animal welfare assurance regarding the beef in the sandwich, (ii) a sandwich with an extra food safety assurance regarding the beef; (iii) a sandwich in which the beef was traceable to the farm of origin; and (iv) a sandwich that combined an animal welfare assurance, a food safety assurance and a traceability assurance. Participants were paid Cdn$ 20 as an incentive for attending the session, and to keep the income endowment constant across participants.

In ten rounds of bidding for each auction sandwich, participants bid the amount they would be willing to pay to exchange their sandwich for the auction sandwich. Individual bids were written down privately by participants and collected as each round of bidding progressed. A Vickrey second-price auction format was used (Shogren et al. 1994). Before starting each bidding round for a specific sandwich, respondents received market information in the form of an announcement of the second-highest bid from the previous round. At the end of the experiment, one round of bidding and one sandwich were selected through a random draw as the binding auction. The highest bidder for the randomly selected sandwich in the selected round of bidding exchanged his/her sandwich for the auction sandwich and paid the second highest bid price. Only one sandwich was auctioned off in each session. The equal chance that any of the rounds of bidding could be binding provides rational participants with the incentive to bid honestly each time.

To evaluate the factors affecting willingness-to-pay (WTP) for traceability and quality assurances in beef, the average WTP bids for each sandwich was regressed against a range of socioeconomic and attitudinal variables using a pooled ordinary least-squares regression model[4]. Three dummy variables represent the different sandwiches from the auction: food safety assurance (MEATSAFETY), animal welfare assurance (HUMANETREAT) and combined traceability, food safety and animal welfare assurances (ALLATTRIBS). The sandwich with a traceability assurance was treated as the reference category. Coefficients on these dummy variables indicate whether respondents were willing to pay a premium over basic traceability for the sandwiches that offered information on specific credence attributes. We expect positive coefficients for the sandwich-related dummy variables if consumers place more value on assurances that reduce information asymmetry with respect to credence (process) attributes, relative to simple traceability.

Three variables measured consumer awareness and concerns over food safety. Direct experience with food poisoning (FPOISON) is expected to induce a higher WTP for additional food safety assurances. Exposure to media coverage of food safety issues (ARTICLES) should impact WTP positively if we assume that these news items were negative. The level of confidence in the current Canadian government food inspection and safety programme (CONFSAFE) is also expected to influence WTP for additional assurances. Given the specification of this variable (see Figure 1), a lower level of confidence in the current food inspection and safety programme is represented by a higher score for the variable CONFSAFE. A positive coefficient would reflect a WTP for stronger (or more reliable) safety and quality assurances than is currently available from the existing food safety inspection system.

Three variables measure the value that respondents said they placed on additional assurances, including assurances about meat safety (VALUESAFE), traceability (VALUETRACE) and on-farm production methods (VALUEPROCESS). These variables correspond to Figures 1-3 and provide a means of checking the validity of the stated preferences for these attributes. As shown in Figures 1-3, a higher rating indicates that the assurance had less value to the respondent. Therefore, we expect these coefficients to be negative if the stated preferences are a good indicator of revealed WTP. The effect of the announced average market price during the first five rounds of bidding was captured with the variable AVEMKTP. The variable is based on the first five rounds of bid data, whereas the dependent variable is based on the last five rounds of bid data to ensure

Table 1. Results of pooled OLS regression analysis (p-values in parentheses)

Variable name	Description	Coefficient estimates
	Dependent variable	
WTP	WTP bids on sandwiches, rounds 6-10	
	Independent variables	
Constant		0.894073*** (0.0000)
HUMANETREAT	Sandwich #1: Humane animal-treatment assurances (Dummy variable)	0.2738471*** (0.0028)
MEATSAFETY	Sandwich #2: Additional food safety assurances (Dummy variable)	0.331288*** (0.0003)
ALLATTRIBS	Sandwich #4: Traceability plus food safety & humane animal-treatment assurances (Dummy)	0.828754*** (0.0000)
FPOISON	Subject or family member experienced food poisoning (Yes = 1)	0.093275 (0.1810)
ARTICLES	News articles/reports read/heard regarding foodborne disease in last 6 months (1 to 7 where 1 = 0-5 articles; 7 = >30 articles)	-0.071054*** (0.0003)
CONFSAFE	Confidence in Canadian food inspection and safety programme (1-5, where 1 = complete confidence; 5 = no confidence)	-0.068306* (0.0708)
VALUESAFE	Value additional assurances about meat safety (1-5, where 1 = highly value; 5 = no value)	-0.128752*** (0.0013)
VALUETRACE	Value knowing exact farm that produced the animal (1-5, where 1 = highly value; 5 = no value)	-0.029843 (0.2980)
VALUEPROCESS	Value knowing processes used by farmer to produce the animal (Score 1-5, where 1 = highly value; 5 = no value)	-0.087259** (0.0203)
AVEMKTP	Average of announced market price from first five rounds	0.074036* (0.0925)
LOCATION	Location of panel (Saskatchewan = 1)	0.281482*** (0.0021)
GENDER	Gender (Male = 1)	0.055950 (0.4405)
AGE	Age (Years)	0.001084 (0.7068)
EDUCATION	Education (1 to 4 where, 1=High school or less; 4 = Graduate degree)	-0.005423 (0.8843)
INCOME	Annual household income (1 to 4 where 1 = under Cdn$30,000; 4 = over Cdn$90,000)	-0.032209 (0.3560)
Adjusted R-squared		0.31320
Number of observations†		412

* = significant at 0.1; ** = significant at 0.05; *** = significant at 0.01

that the market price is exogenously determined with respect to the dependent variable. This variable isolates any market feedback effects from the announced market price, which may indicate strategic bidding on the part of the auction participants. The effect of location was isolated with a dummy variable to distinguish any differences in bidding behaviour between Saskatchewan and Ontario. Four demographic variables are included: gender, age, education and income level. There are no *a priori* strong expectations regarding the effect of these variables on the bids for the four sandwiches. It is unlikely that household income would have a major effect, given the nature of the experiment and the common income endowments with which participants began the experiment.

A cursory analysis at the bid data that form the basis of the dependent variable reveals that traceability to the farm of origin, without additional quality assurances, elicited the lowest average WTP (7% of base sandwich value for beef)[5], and the largest number of zero bids (45%). Quality verification with respect to credence attributes, such as an additional food safety assurance or an animal welfare assurance, elicited higher bids on average[6]. The fourth sandwich, which bundled traceability information with positive quality assurances yielded the highest bids (40%). Due to the nature of a one day experiment, the bid information is usually considered to be an upper bound on WTP (Hayes et al. 1995; Dickinson and Bailey 2002).

Table 1 reports the results of the regression analysis. The coefficients for the three sandwich dummy variables MEATSAFETY, HUMANETREAT and ALLATTRIBS were all significant at 1%. Consistent with *a priori* expectations, the results suggest that a beef sandwich with an extra food safety assurance, or with a humane animal-treatment assurance, could command a premium over beef that was only traceable. Bundling traceability with both of these quality assurances yielded a considerably larger premium over the traceability-only sandwich.

People who said they placed more value on additional food safety (VALUESAFE) and production method assurances (VALUEPROCESS) were actually willing to pay more for the reference sandwich (traceability only) in the beef experiments, verifying the information presented in Figures 1 and 3. Interestingly, this was not the case for people who indicated that they would pay more for a traceability assurance (VALUETRACE). Consistent with the economic functions of traceability discussed earlier, whether people say they value traceability appears to have less of an influence on their actual WTP than an interest in tangible quality assurances with respect to food safety and animal welfare. Traceability information, although helping mitigate the costs of a food safety problem, does not significantly reduce information asymmetry for consumers. The positive and highly significant coefficient on LOCATION implies that Saskatchewan respondents were willing to pay more than Ontario respondents for a sandwich with additional verifiable characteristics.

The negative coefficient for ARTICLES was unexpected and indicates that the more news articles consumers had read about foodborne diseases in the previous six months, the lower their bids for the sandwiches with the verifiable information. This may indicate that news items had reassured consumers, or perhaps that they were sceptical of the information in media articles. Prior experience with food poisoning, and the level of confidence in the Canadian regulatory food safety system, were not significant determinants of WTP. This may explain the apparent inconsistency in Figure 1 between confidence in the food safety system and valuing additional food safety assurances. Even if consumers are confident in the regulatory system, it appears that they may still value additional assurances that offer other assurances with respect to food safety.

The coefficient for average market price was positive and significant at 10%, indicating that there may be limited market feedback effects in the WTP data that are isolated by this variable (Dickinson and Bailey 2002). The remaining variables, including the demographic variables, AGE, GENDER, EDUCATION and INCOME were not statistically significant.

IMPLICATIONS

As this paper has indicated, there are many different types of traceability system emerging as a result of regulatory intervention, at an industry-wide level or as a competitive strategy at the level of individual supply chains. These developments are often prefaced on the underlying assumption that consumers want more traceability. Previously there has been little economic research to evaluate the validity of this assumption, and to assess the extent to which simple traceability delivers benefits to consumers.

The experimental auction methodology used in this study presents a powerful and flexible tool for evaluating consumer preferences for credence attributes. Firms can structure their supply-chain relationships so as to deliver those credence attributes that are valued by consumers. The experimental auction methodology also enables researchers to test the consistency of stated preference attitudes against those revealed through bidding behaviour.

The empirical analysis shows that consumers were willing to pay non-trivial amounts for a traceability assurance, as indicated by the statistical significance of the constant in the regression results. For some consumers, this may imply that they have more confidence in a food product backed by a traceability assurance or, in the event of a problem, that they value the ability to trace products back to source. However, the positive coefficients for the sandwich dummy variables imply that quality assurances with respect to food safety and on-farm production methods for beef were significantly more valuable than a simple traceability assurance. For consumers, traceability has the most value when bundled with additional quality assurances. This finding is consistent with the earlier discussions regarding the functions of a traceability system.

Since these consumer experiments were undertaken, Canada has experienced a few cases of Bovine Spongiform Encephalopathy (BSE), which have been shown to originate in domestic cattle. While these initial BSE cases do not appear to have weakened domestic consumer confidence in the beef industry, it is plausible that a repetition of this experiment post-BSE would reveal higher values for traceability and on-farm production method assurances. Certainly, these issues appear to be garnering closer attention among beef industry stakeholders in the wake of the first domestic Canadian cases of BSE.

Simple trace-back systems are important in limiting the costs from a food safety problem, in maintaining consumer confidence in an industry, and in enforcing liability incentives for due-diligence behaviour. To date, the development of private-sector traceability systems in livestock sectors has primarily been driven by cost- and risk-reduction motivations. While traceability systems can provide the infrastructure to facilitate positive quality assurances, they do not necessarily provide consumers with this additional information. Traceability by itself does not address the issue of consumer information asymmetry with respect to credence quality attributes. As food firms seek to differentiate their products to gain a competitive advantage, bundling traceability with positive quality assurances within a closely monitored supply-chain environment can be the source of future competitive advantage. In this respect, a traceability capability may signal the credibility of quality assurances. The economic rents potentially available from bundling a product differentiation strategy with traceability may be the benefit necessary to offset industry concerns regarding the liability implications of traceability.

NOTES

[1] Article 18 "Traceability". In "Regulation (EC) No. 178/2002 of the European Parliament and of the Council as of January 2002 laying down the general requirements of the food law, establishing the European Food Safety Authority and laying down procedures in matters of food safety". Official Journal of the European Communities.

[2] In an identical set of experiments at the same locations but with 100 different participants, willingness-to-pay for pork product attributes was also assessed. This paper presents only the results from the beef experiments and all statistics refer only to the beef panels. Interested readers are directed to Hobbs et al. (2005) for information and results from the pork experiments.

[3] For a more complete discussion of the experimental auction methodology, model and results, see Hobbs et al. 2005.

[4] The dependent variable was based only on the final five rounds of bidding for a given subject. Data from the first five bidding rounds can be affected by misunderstanding of the auction process, whereas we assume that by the 6th round the bids will have stabilized around a participant's true marginal WTP (Shogren et al. 1994; Hayes et al. 1995).

[5] The average is based on the last 5 rounds of bidding, and is the marginal bid as a percentage of base sandwich value of Cdn$ 2.82 for the beef sandwich. The base sandwich value was calculated by asking respondents how much they would typically expect to pay for this type of sandwich and averaging the responses.

[6] Average willingness-to-pay for a beef sandwich with an additional food safety assurance was 20%, while an animal welfare assurance elicited an average WTP of 18% over the base sandwich value.

REFERENCES

Beef USA, 2005. *National Animal Identification System (NAIS): industry proposal white paper for consideration.* National Cattlemen's Beef Association. [http://hill.beef.org/pdfs/NationalAnimalIdentificationSystemFeb200511.pdf].

Boyer, M. and Porrini, D., 2002. The choice of instruments for environmental policy: liability or regulation? *In:* Zerbe, R.O. and Swanson, T. eds. *An introduction to the law and economics of environmental policy: issues in institutional design.* Elsevier Science, 245-267. Research in Law and Economics Series no. 20.

Buhr, B.L., 2002. *Traceability, trade and COOL: lessons form the EU meat and poultry industry*. International Agricultural Trade Research Consortium, Pullman. Working Paper International Agricultural Trade Research Consortium no. 03-5. [http://agecon.lib.umn.edu/cgi-bin/pdf_view.pl?paperid=8538&ftype=.pdf].

Buzby, J.C. and Frenzen, P.D., 1999. Food safety and product liability. *Food Policy,* 24 (6), 637-651.

Clemens, R., 2003. *Meat traceability and consumer assurance in Japan*. Midwest Agribusiness Trade and Information Center, Ames. MATRIC Briefing Paper no. 03-MPB 5. [http://www.card.iastate.edu/publications/DBS/PDFFiles/03mbp5.pdf].

Dickinson, D.L. and Bailey, D., 2002. Meat traceability: are US consumers willing to pay for it? *Journal of Agricultural and Resource Economics,* 27 (2), 348-364.

Fearne, A., 1998. The evolution of partnerships in the meat supply chain: insights from the British beef industry. *Supply Chain Management,* 3 (4), 214-231. [http://www.imperial.ac.uk/agriculturalsciences/cfcr/pdfdoc/evolution-of-partnerships.pdf].

Fox, J.A., Hayes, D.J., Kliebenstein, J.B., et al. 1994. Consumer acceptibility of milk from cows treated with bovine somatotropin. *Journal of Dairy Science,* 77 (3), 703-707.

Golan, E., Krissoff, B., Kuchler, B., et al. 2003. Traceability for food safety and quality assurance: mandatory systems miss the mark. *Current Agriculture, Food and Resource Issues* (4), 27-35. [http://cafri.usask.ca/j_pdfs/golan4-1.pdf].

Hayes, D.J., Shogren, J.F., Shin, S.Y., et al. 1995. Valuing food safety in experimental auction markets. *American Journal of Agricultural Economics,* 77 (1), 40-53.

Hobbs, J.E., 2003. Traceability in meat supply chains. *Current Agriculture, Food and Resource Issues* (4), 36-49. [http://cafri.usask.ca/j_pdfs/hobbs4-1.pdf].

Hobbs, J.E., Bailey, D.V., Dickinson, D.L., et al. 2005. Traceability in the Canadian red meat sector: do consumers care? *Canadian Journal of Agricultural Economics,* 53 (1), 47-65.

Hobbs, J.E., Fearne, A. and Spriggs, J., 2002. Incentive structures for food safety and quality assurance: an international comparison. *Food Control,* 13 (2), 77-81.

Hobbs, J.E. and Kerr, W.A., 1992. The cost of monitoring food safety and vertical coordination in agribusiness: what can be learned from the British Food Safety Act 1990? *Agribusiness,* 8 (6), 575-584.

MacDonald, J.M. and Crutchfield, S., 1997. Modeling the costs of food safety regulation. *In:* Caswell, J. and Cotterill, R.W. eds. *Strategy and policy in the food system: emerging issues*. University of Connecticut, Amherst, 217-223.

Maple Leaf Foods Inc., 2004. *Maple Leaf Foods Inc.: frequently asked questions*. Maple Leaf Foods Inc. Accessed March 16, 2005

Meat and Livestock Australia, 2005. *National Livestock Identification System: Australia's system for livestock identification and traceability*. Available: [http://www.mla.com.au/content.cfm?sid=131] (March 16, 2005).

Mojduszka, E.M., 2004. *Private and public food safety control mechanisms: interdependence and effectiveness: paper presented at the 2004 American Agricultural Economics Association annual meetings, Denver, Colorado*. [http://agecon.lib.umn.edu/cgi-bin/pdf_view.pl?paperid=14536&ftype=.pdf]

Shogren, J.F., Shin, S.Y., Hayes, D.J., et al. 1994. Resolving differences in willingness to pay and willingness to accept. *American Economic Review,* 84 (1), 255-270.

Souza-Monteiro, D.M. and Caswell, J.A., 2004. *The economics of implementing traceability in beef supply chains: trends in major producing and trading countries*. University of Massachusetts, Amherst. Working Paper Department of Resource Economics, University of Massachusetts no. 2004-6. [http://www.umass.edu/resec/workingpapers/resecworkingpaper2004-6.pdf].

Wasylyniuk, C.R., Bessel, K.M., Kerr, W.A., et al. 2003. *The evolving international trade regime for food safety and environmental standards: potential opportunities and constraints for Saskatchewan's beef feedlot industry*. Estey Centre for Law and Economics in International Trade, Saskatoon. [http://www.esteycentre.ca/BeefFinalReport.pdf].

CHAPTER 8

ON THE (DIS)ABILITY OF THE FIRM TO QUANTIFY CHAINS

A marketing perspective on sharing financial rewards

PAUL INGENBLEEK

Wageningen University, Marketing and Consumer Behaviour Group, Hollandseweg 1, 6706 KN Wageningen, The Netherlands. E-mail: Paul.Ingenbleek@wur.nl.

Abstract. Although the marketing discipline originates from agricultural economics, it currently moves to a new logic that is marked by, among other things, customer value, customer satisfaction, relationships, market orientation and resource-based theories. This article uses this evolving logic in marketing to examine the problem of sharing financial rewards in agricultural supply chains. Building on resource-advantage theory it is suggested that the potential reward that firms may derive from participating in a supply chain depends on the competitive position of the chain as a whole and on the competitive position of the individual firm within the chain. To understand what its contribution to the chain is worth, the firm should be able to quantify relative customer value. The paper identifies inter- and intra-organizational barriers that may disable the firm to do so. Inappropriate assessments lead to a disability of the firm to take financial rewards in exchange for its contribution to the chain. It is questioned whether academicians currently provide chain practitioners with the appropriate approaches to deal with this problem.
Keywords: competition; marketing; pricing; resource-based theory

INTRODUCTION

Having its roots in agricultural economics, the marketing discipline for a long time had a vocabulary and assumptions comparable to those of agricultural economics. Over the last decades, however, marketing is "moving towards a new dominant logic" (Vargo and Lusch 2004, p. 1), which provides an interesting avenue to understand the challenges that agricultural chains nowadays see themselves confronted with. In particular, the number of agricultural chains that differentiate themselves from mainstream production by offering unique products to the consumer seems to increase. These chains have set themselves apart from mainstream production, create more value than their competitors do, and may also have to search for new ways to share the financial rewards for the creation of customer value.

C.J.M. Ondersteijn et al. (eds.), Quantifying the agri-food supply chain, 103 - 115.

In a perfect market, rewards for economic behaviour are determined by the price mechanism, whereas in hierarchies the principal determines the payments to the agent. In a supply chain that is embedded in a network of competing firms, however, firms are torn between the options of competition and collaboration. Relationships emerge when two parties recognize that they both benefit more from exchange within the context of a relationship than from different types of transactions or from transactions with different partners (Anderson and Narus 1984; 1990; Dwyer et al. 1987). The extent to which firms will extract financial rewards from relationships within an integrated supply chain will strongly determine their willingness to participate in that chain. The alternative to a satisfying solution for all participants on how financial rewards are divided would be that some of the participants share the costs while others reap the gains. This may go at the expense of the motivation and income of those participants that share the costs, and it may finally lead to a disintegration of the system.

The remainder of this chapter is structured as follows. The next section will provide some background information on the new dominant logic in marketing. Next, the resource-advantage theory of competition, a cornerstone of this new dominant logic, is described. This theory provides a basis for the subsequent argumentation. First, it is argued that firms compete both *with* and *within* their chain (the topic of the third section). This leads, in the fourth section, to an understanding of the potential reward for contributing to a supply chain. This *potential* reward may, however, deviate from the *actual* reward that the firm receives. If firms are incapable of quantifying the appropriate indicators within their chain (in particular relative customer value), the actual reward may strongly deviate from the potential reward. The fifth section discusses several barriers that may inhibit a firm to quantify relative customer value and thus to cash the full potential reward. The chapter finishes with some conclusions.

THE EVOLVING NEW DOMINANT LOGIC IN MARKETING

The new dominant logic in marketing is marked by concepts like customer value (Woodruff 1997), customer satisfaction (Oliver 1997) and relationship marketing (Dwyer et al. 1987; Morgan and Hunt 1994). A driving force behind these developments has been the field of services marketing (cf. Berry and Parasuraman 1991). The mainstream economic vocabulary appeared to be of little use to the marketing of services. However, according to Vargo and Lusch (2004) services marketing is not the exception but the rule, because in every transaction services are exchanged even if it concerns a transaction between a physical product and a monetary payment. To this respect, a farmer does not just sell his crops: he sells a 'service' by bringing together resources and developing knowledge and the ability to grow crops. This 'service' enables the customer to focus on his own capabilities and deliver services to others. According to Vargo and Lusch (2004) the physical product and the monetary payment (which shape the actual transaction according to economists) 'mask' the actual exchange of services.

Similar to these developments, marketing-strategy literature has shifted its focus from strategy content – like studies on the Profit Index of Marketing Strategies (Buzzell and Gale 1987) – to the resources of firms on which these strategies build. A central concept in the marketing-strategy literature that builds on the resource-based view of the firm (e.g. Dierckx and Cool 1989; Penrose 1959; Wernerfelt 1984) is market orientation. Market orientation refers to the organization-wide generation, dissemination and use of market information pertaining to current and potential customers and competitors (Kohli and Jaworski 1990). Because it is rooted in an organizational culture, market orientation is a resource (Homburg and Pflesser 2000). This resource is leveraged in business processes like strategy-making, new product development and service delivery (Day 1994), which are therefore executed by the organization in ways that lead to the creation of superior customer value (Slater 1997). The creation of customer value subsequently leads to customer satisfaction, customer retention, attraction of new customers (Woodruff 1997) and in the end financial performance (see Rodriguez Cano et al. (2004) for a meta-analysis of relationships between market orientation and business performance).

An important hallmark in the development of the new dominant logic in marketing, are the works of Hunt and Morgan (1995; 1996; 1997) on resource-advantage (R-A) theory. Sharing similarities with many research traditions

Table 1. *Foundational premises of perfect competition and resource-advantage theory (derived from Hunt and Morgan (1997))*

		Perfect competition theory	**Resource-advantage theory**
P 1	Demand:	Heterogeneous across industries, homogeneous within industries, and static	Heterogeneous across industries, heterogeneous within industries, and dynamic
P 2	Consumer information:	Perfect and costless	Imperfect and costly
P 3	Human motivation:	Self-interest maximization	Constrained self-interest seeking
P 4	The firm's objective:	Profit maximization	Superior financial performance
P 5	The firm's information:	Perfect and costless	Imperfect and costly
P 6	The firm's resources:	Capital, labour and land	Financial, physical, legal, human, organizational, informational and relational
P 7	Resource characteristics:	Homogeneous and perfectly mobile	Heterogeneous and imperfectly mobile
P 8	The role of management:	To determine quantity and implement production function	To recognize, understand, create, select, implement and modify strategies
P 9	Competitive dynamics:	Equilibrium-seeking, with innovation exogenous	Disequilibrium-provoking, with innovation endogenous

that deviate from perfect competition theory, R-A theory should be seen as a theory in development, with the final goal to develop into a general theory of competition (Hunt 2000). R-A theory has formulated foundational premises that are closer to actual business practice than those of perfect competition theory (see Table 1). Therefore, it has formulated a theoretical structure on competition that is appealing to both academicians and business people, and that provides a helpful perspective to understand how firms in supply chains share the financial rewards generated by the chain.

RESOURCE-ADVANTAGE THEORY

R-A theory can be explained on the basis of Figures 1 and 2. According to R-A theory, organizations strive to achieve superior financial performance, which can be achieved through a market position of competitive advantage. A position of competitive advantage is a consequence of an organization's advantage in resources compared to competitors (Figure 1). Superior financial performance is "a level of financial performance that exceeds that of its referents, often its closest competitors" (Hunt and Morgan 1995, p. 6). Firms do not maximize profits because they generally lack the information to do so.

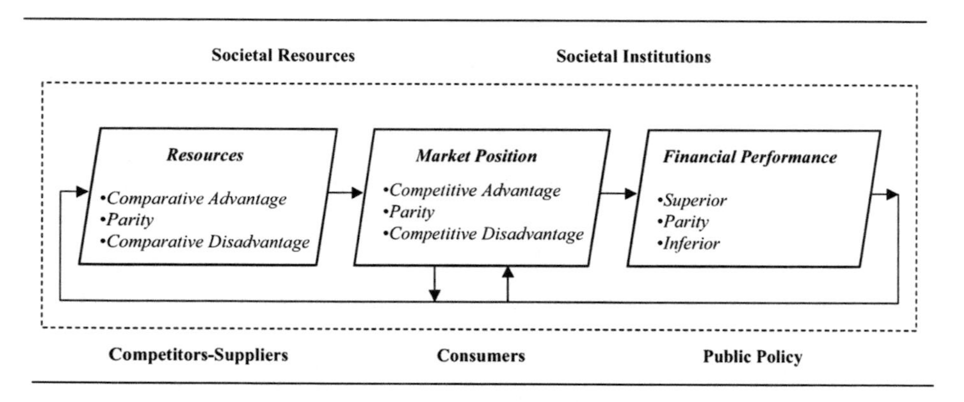

Figure 1. *Resource-advantage competition (derived from Hunt and Morgan (1997))*

Market positions depend on the value that the firm creates to a certain market or market segment on the basis of its resources compared to competitors, as well as on the relative costs that the deployment of resources brings about (Figure 2). Market segments "are intra-industry groups of consumers whose tastes and preferences for an industry's output are relatively homogeneous" (Hunt 2000, p. 11). This suggests that organizations do not compete necessarily within certain industries, but do compete necessarily on certain markets or market segments. Value "refers to the sum total of all benefits that customers perceive they will receive if they accept a particular firm's market offering" (Hunt 2000, p. 32). "*Relative superior value* therefore, equates with *perceived to be worth more*" (Hunt 2000, p. 32, italics in original). This suggests that it is the customer who decides how valuable a market offering is.

		Relative Resource-Produced Value		
		Lower	Parity	Superior
Relative Resource Costs	Lower	1 Indeterminate Position	2 Competitive Advantage	3 Competitive Advantage
	Parity	4 Competitive Disadvantage	5 Parity Position	6 Competitive Advantage
	Higher	7 Competitive Disadvantage	8 Competitive Disadvantage	9 Indeterminate Position

Read: The marketplace position of competitive advantage identified as Cell 3 results from the firm, relative to its competitors, having a resource assortment that enables it to produce an offering for some market segment(s) that (a) is perceived to be of superior value and (b) is produced at lower costs.

Derived from Hunt and Morgan (1997)

Figure 2. *Competitive Position Matrix*

Firms achieve a position of competitive advantage if they create superior value at costs lower than, or equal to their competitors' (cells 3 and 6 in Figure 2, respectively), or if they create value equal to competitors at lower costs (cell 2). In other words: to capture a position of competitive advantage, a firm needs a comparative advantage in its resources that enables it to produce more effectively and/or efficiently than its competitors. A firm obtains a position of competitive disadvantage if it creates relatively lower value at costs equal to or higher than their competitors (cells 4 and 7), or if it creates value equal to their competitors' at higher costs (cell 8). Cell 5 represents a parity position. In this situation, all firms competing on a certain market or market segment have relatively equal resource-produced value and relatively equal resource costs. A firm that occupies a market position represented by cell 1, in which it creates lower value at lower costs, will have to set lower prices than competitors in order to have a chance at achieving competitive advantage. Also if the firm creates relatively higher value at relatively higher costs, its position is indeterminate (cell 9). Its competitive advantage depends here on the willingness of customers to pay premium prices in return for market offerings of superior value.

The process of R-A competition is dynamic. In order to achieve a position of competitive advantage, firms continuously seek for a comparative advantage in resources. R-A theory defines resources as: "the tangible and intangible entities available to the firm that enable it to produce efficiently and/or effectively a market offering that has value to some market segment(s)" (Hunt 2000, p. 11). Resources are of various kinds in R-A theory: financial, physical, legal, human, organizational, informational and relational. Resources may be the result of the firm's past and they may be imperfectly mobile, such as relationships with customers and suppliers. Achieving superior financial performance enables the firm to invest in resources. Firms can improve their market positions by introducing innovations to the market. As such, competitive positions are not stable. Positions of competitive advantage can be sustained if competitors base them on resources that are difficult to imitate or obtain.

Firms may learn from the process of competition. If the firm achieves a certain degree of performance, it may learn about the competitive position and the specific resources on which this position is based. By learning from the process of competition, a firm may learn in which resources it should invest in order to improve its position. Considering that a firm may learn the wrong things, a position can be harmed if the firm invests in the resources that do not lead to a position of competitive advantage.

Customers, competitors, suppliers, societal institutions, public policy and societal resources influence the process of R-A competition. Customers' preferences may change, competitors may imitate certain types of resources, suppliers may raise their prices, etc. These stakeholders may impact on the comparative advantage of resources as well as on the explicit and implicit 'rules of the game'. Societal resources impact on the firm's resources, like the availability of natural resources such as oil, or the level of education in a society. Resources of a legal nature, like patents, may protect innovations, while environmental or safety laws may force firms to modify production plants and processes.

COMPETITION WITHIN AND BETWEEN CHAINS

In R-A terms chains strive for a comparative advantage in resources, which results in a position of competitive advantage in a certain market or market segment, which yields superior financial performance to the chain as a whole. Chains are clear examples of how relational resources may work out: together chain members may compete more effectively and/or efficiently than they might individually. Competition between chains is depicted in Figure 3. Both chains compete in the same market segment, trying to be more effective and/or efficient than the competing chain.

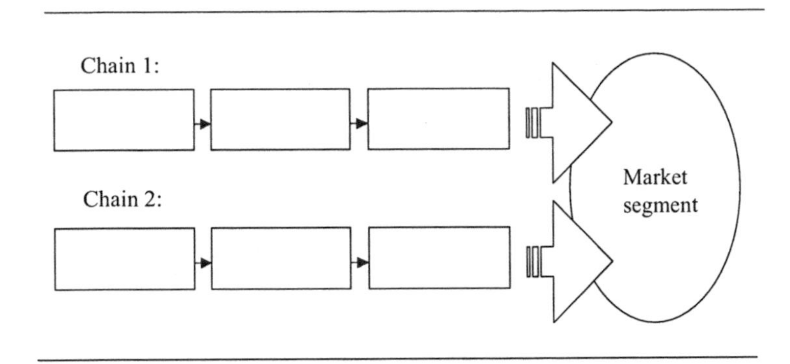

Figure 3. Competition between chains

However, in order to achieve superior financial performance, firms do not compete with their chain against other chains alone, they also compete within their chain (see Figure 4). In the continuous struggle for a comparative advantage in resources, chains may improve their stock of resources by involving new partners in the chain and removing others. Firms that possess resources that make the chain compete more effectively and/or efficiently, may enter the chain at the expense of others. Firms may participate in multiple chains, strategically deploy resources over them, scan the environment for new opportunities, assess the importance of current relationships, and assess the potential of new ones. If the firm has a strong resource stock, it can easily switch (think of the powerful positions that many food retailers occupy). However, if relationships are strong, it is unlikely that firms are quickly removed from a supply chain when they find themselves in a position of competitive disadvantage. Instead, chain partners are more likely to allow them some time to strengthen their positions (Morgan and Hunt 1994).

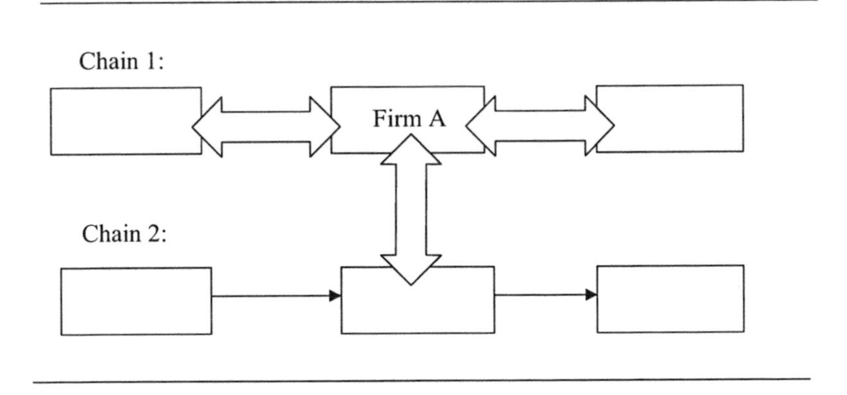

Figure 4. Competition within a chain

In sum, both competition *between* chains and competition *within* a chain determine firms' financial performance. The frameworks of R-A competition can be applied both to the firm and to the chain. Firms compete within these two – sometimes-conflicting – systems.

THE POTENTIAL REWARD FOR CONTRIBUTIONS TO A SUPPLY CHAIN

Considering that the financial performance of a chain should be divided over its members, the reward that a firm will receive for participating in a chain is some share of the financial performance of the chain as a whole. We can write the firm's reward for chain activities (R) therefore as:

$$R = f\,[share, financial\ performance_{chain}]. \quad (1)$$

In order to reward a firm for its contribution to the financial performance of the chain, the ratio of distribution should be based on the firm's market position. If the firm deploys a comparative advantage of resources in the chain, this contributes to the market position of the chain. Rewarding firms on the basis of their market position strengthens the relationships between resources and market positions and between market positions and financial performance, thereby speeding up the process of R-A competition, productivity and economic growth. Given that the ratio of distribution is in reality often not entirely based on the firm's market position, we speak of a normative function in which we try to explain the firm's potential reward for chain activities (PR) rather than its actual reward. The potential reward is the maximum amount of money that a firm may extract based on its contribution to the chain.

$$\text{PR} = f\,[\text{market position}_{\text{firm}}, \text{financial performance}_{\text{chain}}]. \quad (2)$$

Given that in R-A theory financial performance is a consequence of a market position, we can replace the financial performance of the chain in this function by the market position of the chain:

$$\text{PR} = f\,[\text{market position}_{\text{firm}}, \text{market position}_{\text{chain}}]. \quad (3)$$

Since relative value and relative costs determine a market position, we can specify the function further. By relative costs is meant the costs of deploying resources in the activities of the chain relative to a perceived alternative. This is an alternative for a firm's activities in a chain, which may be either a competitor, forward integration, backward integration, or a network extension. Relative value is the sum total of all benefits that the next link in the chain perceives it will receive from chain collaboration relative to a perceived alternative (based on Hunt 2000, p. 32).

Costs represent the lower boundary: the minimum amount the firm should receive for enabling its resources in the chain without making a loss. Value represents the upper boundary: what the result of deploying resources is worth to the customer (see also Figure 5). The potential reward for deploying resources is therefore a function of the relative value (RV) created by the firm minus its relative costs (RC) of enabling resources, and the relative value created by the chain to the target market (segment) minus the costs of enabling the resources of the chain:

$$PR = f[(RV_{firm} - RC_{firm}), (RV_{chain} - RC_{chain})]. \tag{4}$$

QUANTIFYING CUSTOMER VALUE

In agricultural chains that differentiate themselves from mainstream production by delivering unique benefits to the consumer, it is essential to assess the upper-boundary, i.e. to quantify relative customer value. Clearly, if the firm uses some proxy to quantify value that is actually much lower than the value perceived by the customer, it grants the customer with a surplus that is higher than necessary.

In an empirical analysis of the effects of firms' pricing practices on profit margins of innovations, Ingenbleek et al. (2004) show that firms that create superior value are often incapable of expressing this value in the price they receive in return. This inability may be caused by a lack of information on a reference point in the market (what do others charge for their products) and/or a lack of information on how much better the firm's innovation is compared to this reference point (see Figure 5).

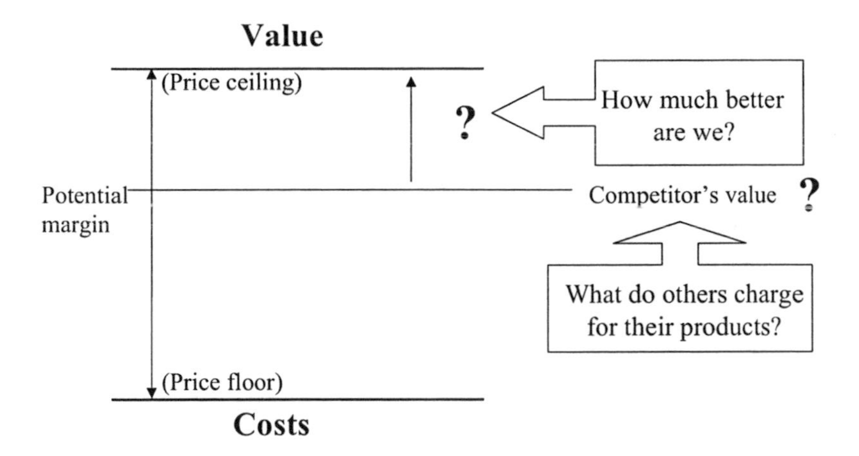

Figure 5. *Quantifying the potential reward (adapted from Monroe 2003)*

Given this process of quantifying relative customer value, firms may bring the actual reward for their chain activities close to the potential reward if two conditions are satisfied. First, they should be able to assess reasonably their position of relative customer value (meaning that they should be able to detect the reference points in the market and to assess how much more value they deliver as compared to these reference points). Second, other chain partners should also be capable of quantifying their relative-value position, because the degree to which downstream chain members are rewarded will determine the extent to which they can possibly pass on these rewards to upstream chain members. In the next section the barriers are discussed that disable firms to assess relative customer value.

BARRIERS TO QUANTIFY RELATIVE CUSTOMER VALUE

A *potential* reward suggests that this reward is not for granted. In fact, it is probably impossible to extract the full potential reward for a firm's activities in a supply chain, because it is virtually impossible to quantify what a market offering is precisely worth to the customer. Information on customer value is ambiguous by definition (Sinkula 1994) and it may be quickly outdated in the dynamic process of R-A competition. In the barriers that prevent firms to take the potential reward for chain activities, we may distinguish between inter-organizational barriers (referring to relationships between firms) and intra-organizational barriers (referring to relationships within firms).

Inter-organizational barriers

Information to assess customer value may be acquired from multiple sources, of which relationships in the chain and its surrounding network are probably the most important (Granovetter 1973; Hansen 1999; Rindfleisch and Moorman 2001). Quantifying customer value requires insight into the customer (Anderson and Narus 1999). In order to assess how much one contributes as an upstream supplier to the market position of a downstream customer, one needs insight into the market position that the customer occupies at his/her customer (Ingenbleek 2004). This requires detailed information from the customer, which can be obtained only if the firm has developed a strong relationship with its customer. Weak relationships within chains, i.e. relationships that lack a sufficient level of trust and commitment (Morgan and Hunt 1994) are therefore the first barrier to quantifying relative customer value.

The second barrier may be a lack of contacts beyond the strong relationships in the chain. A drawback of strong relationships may be that they have a blinding effect on actors (Granovetter 1973). As indicated in Figure 5, to assess relative customer value, firms need to be able to assess reference points, which are most likely their closest competitors. In order to keep track of these competitors and possible new entrants, firms should not concentrate too much on the relationships within their own chain, but stay in business (through weaker relationships than the relationship in their major chain) with others that can provide such information (Ingenbleek 2004).

Third, because innovative means of value creation often require innovative price mechanisms that determine the rewards for the created customer value, the existing price mechanisms within chains may be a barrier. There are often well-established ways on how prices (rewards) are calculated or determined in chains, laid down in contracts, routines or perhaps institutions such as auctions. If these established price mechanisms are not based on a quantification of relative customer value and if actors are reluctant to switch to new price mechanisms, they are a barrier.

Intra-organizational barriers

If the firm has established the appropriate network contacts and thus can acquire the appropriate information, several barriers within the firm may inhibit its ability to quantify relative customer value. A first barrier may be the transmission of the information within the firm (Huber 1991; Maltz and Kohli 1996). Information should be transmitted to those who are responsible for the price decision or negotiations with customers. In order to enable chain partners to increase their rewards, firms should also reward upstream partners for their contribution to the chain. This requires information to be transmitted to purchasers.

Second, managers should not just have the information; they should also interpret it correctly. In order to use both customer and competitor information in price decisions, managers need interpretation schemes that are rooted in a market-oriented culture (Day and Nedungadi 1994). In organizations there may exist tendencies to use other types of information in order to avoid the ambiguity of value information (Adams et al. 1998). When weighted against less ambiguous information such as price discounts, purchasers put less weight on value information (Anderson et al. 2000).

Third, even if information on relative value is acquired, distributed to the relevant business functions and correctly interpreted, it may not always be used in decision-making. Management systems should be aligned with the firm's objectives of value creation. If the firm rewards its sales people for market share rather than profits, and its purchasers for cost-cutting rather than value increases, these managers are unlikely to use the information on relative value (cf. Ingenbleek and De Vlieger 2004).

CONCLUSIONS

To ensure that chain members remain motivated to invest in the chain and to provide them with sufficient financial resources to do so, it is in the common interest of all chain members that each of them is rewarded for its contribution to the competitive position of the chain. It is also in the best interest of public policy, if public policy aims for economic growth. In other words: the actual reward for the chain members' contributions to the chain should be as close as possible to the potential reward. Chain members should both 'live and let live': cash the rewards for their own contribution to the market position of the chain and allow other chain members to take a share based on their contribution. To this respect, the view of pricing as a capability is endorsed here. As Dutta et al. (2003, p. 629) suggest: "Managers in a firm without effective pricing processes may be unable to set prices that reflect the wishes of their customers, so the customers may misuse resources. As such effects ripple through a supply chain or a market sector, society may be worse off because resources are used inefficiently".

The capability that enables firms to take the rewards for their activities in a chain, is fed by both competitor and customer information. These types of information enable it to assess relative customer value. In order to collect these types of information, distribute them to the appropriate business functions, interpret them correctly and use them in actual decision-making, firms may see themselves confronted with barriers that exist within their own firm and between their firm and their chain partners.

The view presented here to clarify the problem of sharing financial rewards in chains, poses an important question to widely used approaches for studying agricultural chains, such as industrial economics and transaction-cost economics: are these approaches still helpful to solve questions on how financial rewards should be divided among chain partners, or should we move to alternative approaches? As agricultural chains increasingly seek to create customer value and differentiate themselves from mainstream production, new approaches based on the new dominant logic in marketing may be promising for the future.

ACKNOWLEDGEMENTS

Paul Ingenbleek is assistant professor in marketing at Wageningen University and scientific researcher at the Agricultural Economics Research Institute (LEI) in The Hague. The author thanks KLICT for sponsoring this research and George Beers, Menno Binnekamp, Gert-Jan Hofstede, Robert Hoste, Joost Krul, Jan-Willem van der Schans and Theo Verhallen for the helpful comments during the realization of this chapter.

REFERENCES

Adams, M.E., Day, G.S. and Dougherty, D., 1998. Enhancing new product development performance: an organizational learning perspective. *Journal of Product Innovation Management,* 15 (5), 403-422.

Anderson, J. and Narus, J., 1990. A model of distributor firm and manufacturer firm working partnerships. *Journal of Marketing,* 54 (January), 42-58.

Anderson, J.C. and Narus, J.A., 1984. A model of the distributor's perspective of the distributor-manufacturer working relationships. *Journal of Marketing,* 48 (Fall), 62-74.

Anderson, J.C. and Narus, J.A., 1999. *Business market management: understanding, creating, and delivering value.* Prentice Hall, Upper Saddle River.

Anderson, J.C., Thomson, J.B.L. and Wynstra, F., 2000. Combining value and price to make purchase decisions in business barkets. *International Journal of Research in Marketing,* 17 (4), 307-329.

Berry, L.L. and Parasuraman, A., 1991. *Marketing services.* Free Press, New York.

Buzzell, R.D. and Gale, B.T., 1987. *The PIMS principles: linking strategy to performance.* Free Press, New York.

Day, G.S., 1994. The capabilities of market-driven organizations. *Journal of Marketing,* 58 (October), 37-52.

Day, G.S. and Nedungadi, P., 1994. Managerial representations of competitive advantage. *Journal of Marketing,* 58 (April), 31-44.

Dierckx, I. and Cool, K., 1989. Asset stock accumulation and sustainability of competitive advantage. *Management Science,* 35 (12), 1504-1511.

Dutta, S., Zbaracki, M.J. and Bergen, M., 2003. Pricing process as a capability: a resource-based perspective. *Strategic Management Journal,* 24 (7), 615-630.

Dwyer, F.R., Schurr, P.H. and Oh, S., 1987. Developing buyer-seller relationships. *Journal of Marketing,* 51, 11-27.

Granovetter, M.S., 1973. The strength of weak ties. *American Journal of Sociology,* 78 (6), 1360-1380.

Hansen, M.T., 1999. The search-transfer problem: the role of weak ties in sharing knowledge across organization subunits. *Administrative Science Quarterly* (March), 82-111.

Homburg, C. and Pflesser, C., 2000. A multiple-layer model of market-oriented organizational culture: measurement issues and performance outcomes. *Journal of Marketing Research,* 37 (4), 449-462.

Huber, G.P., 1991. Organizational learning: the contributing processes and the literatures. *Organization Science,* 2 (1), 88-115.

Hunt, S.D., 2000. *A general theory of competition: resources, competences, productivity, economic growth.* Sage, Thousand Oaks.

Hunt, S.D. and Morgan, R.M., 1995. The comparative advantage theory of competition. *Journal of Marketing,* 59 (April), 1-15.

Hunt, S.D. and Morgan, R.M., 1996. The resource-advantage theory of competition: dynamics, path dependencies, and evolutionary dimensions. *Journal of Marketing,* 60 (October), 107-114.

Hunt, S.D. and Morgan, R.M., 1997. Resource-advantage theory of competition: a snake swallowing its tail or a general theory of competition? *Journal of Marketing,* 61 (4), 74-82.

Ingenbleek, P., 2004. Pricing in order to increase profit margins in marketing channels: towards an integration of pricing with resource-based and social network theories. *In:* Verhallen, T. and Gaakeer, C. eds. *Demand driven chains and networks.* Reed Business Information, 's-Gravenhage, 139-160.

Ingenbleek, P. and De Vlieger, K., 2004. Het waarderen van ketenpartners in kwaliteitsgerichte ketens: Hoezo Waardering? *VMT* (12), 54-55.

Ingenbleek, P., Frambach, R.T. and Verhallen, T.M.M., 2004. *Pricing to increase profit margins: an empirical examinations of firms' pricing practices: paper presented at the EMAC Conference, Murcia, May 2004.*

Kohli, A.K. and Jaworski, B.J., 1990. Market orientation: the construct, research propositions and managerial implications. *Journal of Marketing,* 54 (April), 1-18.

Maltz, E. and Kohli, A.K., 1996. Market intelligence dissemination across functional boundaries. *Journal of Marketing Research,* 33 (1), 47-61.

Monroe, K.B., 2003. *Pricing: making profitable decisions.* McGraw-Hill/Irwin, Boston. Oorspr. uitg.: 1979

Morgan, R. and Hunt, S., 1994. The commitment-trust theory of relationship marketing. *Journal of Marketing,* 58 (3), 20-38.

Oliver, R.L., 1997. *Satisfaction: a behavioral perspective on the consumer.* MacGraw-Hill, New York.

Penrose, E.T., 1959. *The theory of the growth of the firm.* Blackwell, Oxford.

Rindfleisch, A. and Moorman, C., 2001. The acquisition and utilization of information in new product alliances: a strength-of-ties perspective. *Journal of Marketing,* 65 (2), 1-18.

Rodriguez Cano, C., Carrillat, F.A. and Jaramillo, F., 2004. A meta-analysis of the relationship between market orientation and business performance: evidence from five continents. *International Journal of Research in Marketing,* 21 (2), 179-200.

Sinkula, J.M., 1994. Market information processing and organizational learning. *Journal of Marketing,* 58 (1), 35-45.

Slater, S.F., 1997. Developing a customer value-based theory of the firm. *Journal of the Academy of Marketing Science,* 25 (2), 162-167.

Vargo, S.L. and Lusch, R.F., 2004. Evolving to a new dominant logic for marketing. *Journal of Marketing,* 68 (1), 1-17.

Wernerfelt, B., 1984. A resource-based view of the firm. *Strategic Management Journal,* 5, 171-180.

Woodruff, R.B., 1997. Customer value: the next source for competitive advantage. *Journal of the Academy of Marketing Science,* 25 (2), 139-153.

MODELLING AGRI-FOOD CHAINS

CHAPTER 9

PROFITABILITY OF 'READY-TO-EAT' STRATEGIES

Towards model-assisted negotiation in a fresh-produce chain

HANS SCHEPERS[#] AND OLAF VAN KOOTEN[##]

[#]Agrotechnology and Food Innovations, Wageningen University and Research Centre, P.O. Box 17, 6700 AA Wageningen, The Netherlands. E-mail: Hans.Schepers@wur.nl

[##]Horticultural Production Chains Group, Wageningen University and Research Centre, Marijkeweg 22, 6709 PG Wageningen, The Netherlands

Abstract. With help of a simple System Dynamics model describing purchase frequency of consumer segments, we illustrate under which circumstances a scenario of 'ready-to-eat' positioning creates value for retailer, trader and grower in a fruit production chain. Although fully quantified, the model is meant as a discussion support tool. It illustrates how collaboration affects the pay-offs of innovation for each trade partner. We show how negotiations addressing other factors than prices optimize total chain profit and hence profit per player. These factors include ready-to-eat positioning, the variation in product ripeness within batches, and cost-sharing agreements regarding product loss and the promotional budget.
Keywords: mathematical modelling; product positioning; consumer behaviour; willingness to innovate; economics of collaboration

INTRODUCTION

Innovation in chains poses dilemmas: taking action by one player influences other players' profitability, or the profitability of other players' collaborative or competitive actions. For example, actions may only be marginally effective unless other players proceed with complementary actions. Improving the quality of perishable produce by the supply chain in order to stimulate consumption is thwarted when the retailer uses the longer shelf life to ship products to more distant locations in order to compete there for new volume, based on less 'fresh' product. Therefore, timely discussions on potentially innovative value-creating options between chain (e.g. trade) partners can maximize total extra profits to the chain players involved, and prevent loss of pay-off of innovations due to intra-chain competition. Such discussions require that the often implicit assumptions necessary for the (financial) implications of the projected innovations are made explicit. For

C.J.M. Ondersteijn et al. (eds.), Quantifying the agri-food supply chain, 119-134.

many operational processes such as logistics, detailed quantitative models are used to calculate new profits and optimal design choices. However, for a product re-positioning (e.g. ready-to-eat fruits), various factors that are difficult to quantify – let alone predict – such as consumer perceptions of product quality and effectiveness of promotions, are vital to address. This adds a substantial amount of uncertainty and leaves much room for misunderstanding, based on differing perspectives, expectations and experiences regarding the effectiveness of possible marketing actions. Our hypothesis is that *especially* in such situations, modelling these assumptions and possible mechanisms behind changed consumer behaviour improves the strategic discussions, and is instrumental in increasing the success rate of the innovations involved. However, in contrast to the applications in well-quantified fields such as logistics, not many tools and basic mental models exist to analyse the potential effect of innovations involving consumer behaviour. Here, we model the effect of ready-to-eat positioning of fresh exotic fruits on the potential consumed *revenues* (in contrast to cost levels) for each player in the chain. Very quickly however, we encounter the main hurdle: product loss. It is due to the combined modelling of product loss and the newly generated revenues that the model can point to viable collaborative chain arrangements that facilitate optimal profit for each player.

Exotic fruit case

Supply chains can minimize product loss of exotic fruits (e.g. mangoes) and stone fruits (e.g. fresh peaches) by harvesting and selling at a quite early (unripe) development stage. However, this also means that consumers should let the fruit ripen at home for a number of days before consuming it. In practice, they may lack the patience to let the fruit properly ripen, and consume it while it does not yet have the taste and texture properties they actually value and expect. As a result, a fraction of disappointed consumers remain in a stage of 'low frequency' usage. Our hypothesis is thus that too strict application of agro-logistics expertise focused on minimization of product loss may negatively impact the volume of consumption.

Clearly, winning consumers for fresh exotic produce is a complex and difficult task. Pricing, promotion and product quality are among the driving forces that increase consumption. Thus, the distribution of fresh but perishable produce has to meet more conditions than just a swift response to the change in quantity of products bought by consumers as for instance 'Efficient Consumer Response' aims to do. Much additional effort is involved in preventing quality loss during storage, transportation and display on the retail shelves. Parameters such as storage timing and especially temperature regimes can be chosen in a way either to minimize product loss, to meet or exceed expected quality at the retail shelve, or to cover a larger geographical distance between the producer's and retailer's locations. In addition, store-front parameters, such as packaging and the display regime (last-in-first-out, first-in-first-out) are important controllable parameters, in order to ensure timely sales of produce before quality loss hampers the attractiveness to consumers. Finally, discounting on the retail price just prior to the use-by-date (explicit or implicit) provides an additional measure. In that case, the variation in freshness within a batch of products is communicated explicitly and used to the advantage of both consumers and retailers.

Here we present a system-dynamics model that should help to explore the effectiveness of the various tactical marketing decisions for product positioning, promotions and pricing. The consumption of mangoes is used as an illustrative example. The model allows one to study incentives of each chain player to benefit from such marketing efforts. For example, sharing cost of promotion between trade partners makes it possible to identify situations where neither player would innovate on its own, while through a cost-sharing agreement they both benefit sufficiently from marketing efforts. By capturing generic dilemmas of post-harvest product handling, marketing and chain collaboration, we can learn how these innovation efforts are best combined. Steady-state analysis and optimization techniques are used to generate optimal innovation policies for retailers and producers.

THE MODEL

The model integrates heuristics from three disciplinary domains, viz., consumer science, quality management and chain management, as illustrated in Figure 1.

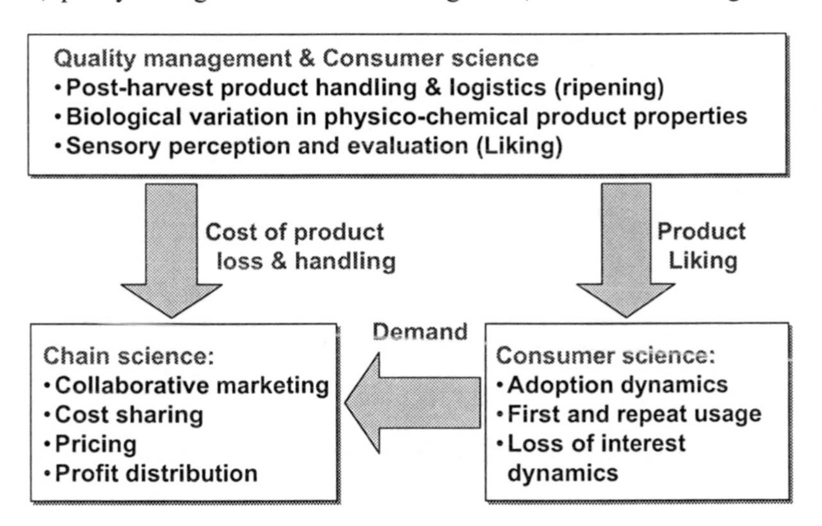

Figure 1. *Three research fields provide components of the model. The linking variables are indicated with arrows*

Quality management aims to provide an attractive product to consumers at minimum cost for product handling and cost for product loss. Consumer liking drives usage dynamics and thus provides 'demand'. Within chain-management science, collaborative marketing, involving smart pricing and cost sharing can yield the right incentives for chain players (in our case the retailer and its trade partner, here called 'producer') to realize a viable business, and ultimately employment, in the exotic-fruits sector.

The time scale of strategic interest to the chain players is assumed to be medium term, i.e. around four years. Since consumer adoption is a dynamic process that takes place on that time scale, it is explicitly translated into differential equations that lie at the heart of the model. Faster processes (e.g. generation of product loss and cost) are translated into auxiliary variables that are determined by other variables and parameters. Slow determinants or processes are represented as constant parameters (e.g. unit handling cost, consumer preferences, price elasticity). For generic system-dynamics modelling as applied to business strategy and practice, see Sterman (2000).

We present the three sub-models in the following paragraphs, and refer to Appendix 1 for the formal mathematical equations.

Modelling consumer-behaviour matters

The sub-model describing consumer behaviour is a first-order compartment model for consumer segments: non-users (N), low-frequency, 'occasional' users (L) and high-frequency, 'loyal, repeat' users (H), shown in Figure 2.

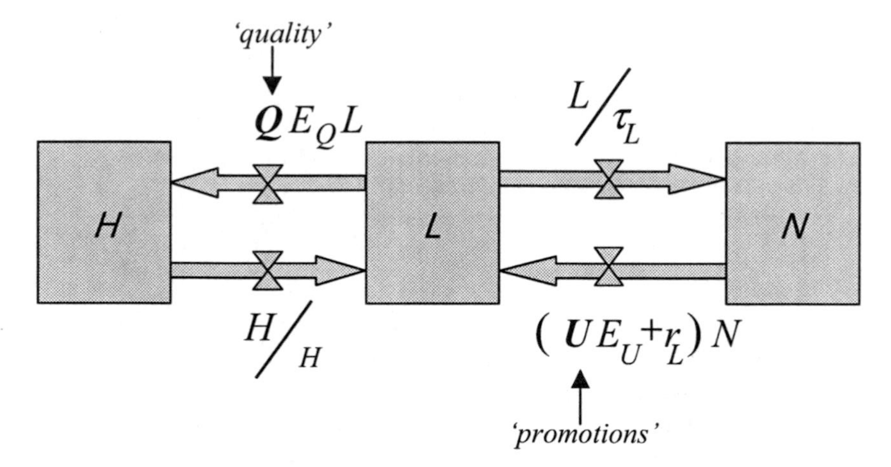

Figure 2. *The consumer sub-model consisting of three product adoption segments*

Promotions (U) and Product quality (Q) modulate different rates of adopting a more frequent product usage. The transition from non-user (N) to occasional user (L) is assumed to occur either spontaneously or under the influence of 'promotions' (U). The development of product loyalty, i.e. the transition from occasional user (L) to repeat user (H) is driven by Liking, which is defined below. When the product does not elicit Liking, occasional users (L) fall back to non-users (N) with a constant rate. For completeness, also loyal consumers may lose interest and 'fall back' into the state of occasional user (L), with a constant rate. How these processes are affected by quality, price and promotions is not included in the present model. Promotion and 'offering the right quality' are the two main instruments to obtain many and loyal customers, respectively. Note that this part of the model describes the intended usage frequency only, not the actual buying frequency, which depends on the price encountered in the store, governed by a standard price-elasticity curve. This 'standard' treatment of the effect of price is included in the computation of volume sold (see Appendix 1).

Modelling quality matters

On the product-handling side of quality management, we allow for a substantial amount of biological variation in the ripeness of fruits at the moment of harvest. For simplicity, this variation remains constant during transportation and storage. Technologies may exist to reduce the variability within a batch, e.g. by sorting. This affects the product-loss rate on the supermarket shelves as well as the Liking, as will be shown below.

Liking is modelled here as the extent to which consumers find what they expect in the product offering. For simplicity, the attribute space spanning relevant product properties was taken as one-dimensional ('ripeness'), which is intended to correspond to the first principal component of a full collection of product attributes. The Liking curve can be thought of as a demand curve: in our case a function of the 'objective' product quality characteristics (i.c., ripeness), instead of a function of price. For exotic fruits, it is assumed to be bell-shaped, with an optimum ripeness and a relatively large region of tolerance for ripeness around it. The Liking curve thus also reflects the various preferences of consumers based on 'optimal taste', combined with the duration consumers may wish to store the fruit at home before consuming it. For various forms of the Liking curve, see Schepers et al. (2004).

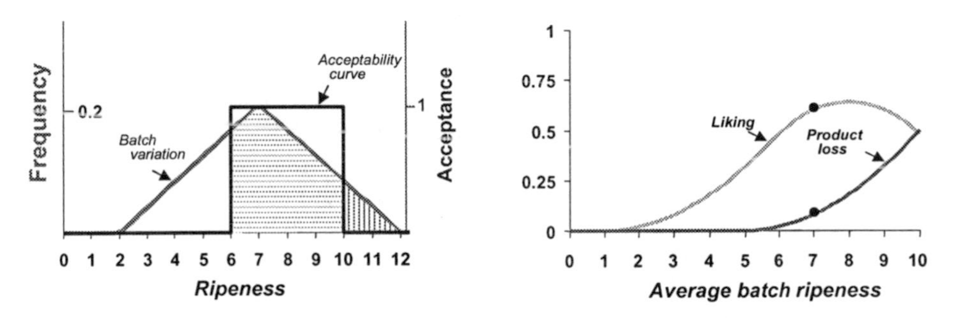

***Figure 3**. Ripeness, Liking and product loss*

Ripeness is represented on an arbitrary scale from 0 (unripe) and 10 (start of over-ripe). Mangoes having a perceived firmness above 10 are considered to be over-ripe and unsuitable for consumption: therefore the starting point for product loss, as the mangoes cannot be sold anymore. In the simplest case, illustrated in Figure 3, consumer liking is modelled with a uniform Liking function ranging with a lower limit (minimum ripeness) of 6 and a maximum acceptable ripeness of 10. In Figure 3 (left-hand panel), an alternative definition of product quality is given by mapping both the distribution of *supplied* product property (ripeness, 'objectively measurable') and *demanded* (expected, appreciated ripeness) product properties ('subjectively perceived attribute') into the relevant attribute space. We define Quality (Q in the Appendix) as the overlap, indicated as the horizontally shaded area; mathematically, the integral of batch variation distribution function and the Liking function. At the same time, the product-loss fraction is automatically defined as the integral (W in the Appendix) of the variation distribution function above the cut-off point (here at 10), indicated by the vertically shaded area. How product loss increases with average batch ripeness is also shown on the right-hand panel. The batch variation causes the product loss to rise when the average ripeness increases.

In order to quantify the cost of providing fruits of various ripeness levels, we assume that the time it takes for the fruits to ripen from one stage to the next is one day, of yet unspecified treatment. More elaborated models for ripening speed as function of temperature and specific technologies, such as using ethylene or various forms of packaging, can be coupled at a later stage.

Modelling chain-behaviour matters

In order to model how chain players can collaborate to provide the right product quality, *given* the preferences and the familiarity of consumers with the product, the sub-models for both the dynamics of the consumer segments and that of the quality perception are combined. Subsequently, we compute the financial implications (e.g. profits Y) of the marketing strategy for each of the three players in the chain: the retailer (index r), the trader or importer (t) and the producer or exporter (p). A subscript (c) denotes the summed profits of the retailer and the trader. Profits are reported with dimensions euros (€) per week, and could represent the profits for one supermarket or outlet (numbers are thus quite low). The profits Y_r, Y_t and Y_p are straightforward functions of prices, the volume and the unit cost level for the producer C_p, for the retailer, handling costs per day C_{hd} and per fruit C_{hp}, and the product-loss fraction W. The equations are described in Appendix 1. For causal diagrams and a Stock & Flow representation of the model, we refer to Schepers et al. (2004). One extension, and implicitly a correction, from Schepers et al. (2004) concerns the allocation of the cost of product loss (the case described in that paper is an illogical special case of the following generic form): we here define the fraction f_w of the cost of the volume that is wasted as product loss the retailer has to pay for (to the trader). In certain product groups, e.g. milk, the retailer returns the unsold products to the manufacturer, who pays back (part of) the volume. In such cases, f_w would be zero or at least below 1, whereas for fresh produce, today, the retailer is not reimbursed for unsold volume, and f_w is 1. Table 1 presents the parameters, symbols and values. The initial conditions are set at the analytically computed steady-state level corresponding to the other parameter constants.

Another chain management instrument already present in the model is the fraction of joint product-promotion budget provided by the trader, u_t. Other parameters on which negotiations/bargaining may take place are the transfer prices P_{rt} and P_{tp} and the amount of variation in ripeness within a weekly batch of fruit, v. This important parameter determines to a large extent the product loss and, as we shall see, the extent to which consumers' expectations are met, resulting in better or worse quality.

Table 1. *Parameter constants and initial conditions*

	Parameter	Symbol	Value	Dimension
Producer/ trader	production unit cost	P_{tp}	0.25	€/piece
	trader share in promotion	u_t	50%	-
	promotion budget (chain cost)	U	65	€/week
Retailer	variation	v	5	days
	freshness deadline	m	10	days
	product positioning	T	5	days
	handling cost per day	C_{hd}	0.005	€/piece/day
	retail fixed handling unit cost	C_{hp}	0.1	€/piece
	purchase price	P_{rt}	0.5	€/piece
	price positioning	d	1	-
Consumer	minimum acceptance	a	6	days
	time scale to stop using ($L \rightarrow N$)	τ_L	26	weeks
	time to lessen consumption ($H \rightarrow L$)	τ_H	52	weeks
	total number of consumers	Z	10000	persons
	initial number of occasional users	$L(0)$	2680	persons
	initial number of repeat users	$H(0)$	669	persons
	consumption occasional user	D_L	0.02	pieces/(week* person)
	consumption repeat user	D_H	0.3	pieces/(week* person)
	quality effect of frequency	E_Q	0.015	1/week
	price elasticity	e	-3	-
	reference price	P_{cr}	1	€/piece
	promotional effectiveness	E_U	0.0002	1/€
	fraction spontaneously trying anew	r_L	0.0025	1/week

RESULTS

We here present a limited set of instruments the producer and retailer have, to influence consumer behaviour and to optimize their economic return. In this exemplary study, the average firmness T is used by the retailer to influence the transition of occasional (L) users to repeat users (H). Both retailer and producer gain a profit from promotion U to enhance the transition of non-users (N) to occasional users (L). But the retailer and producer may bargain about their share of the promotion costs they contribute. Equally, the share of the product-loss cost (f_w) (reimbursement) will be shown to modulate the incentives to proceed on adjusting other important parameters, notably the variance in ripeness within each weekly batch, which is most relevant for a profitable ready-to-eat strategy.

The main use of the model in its current unvalidated state is as discussion support tool, aiming at discussing qualitative responses to qualitative changes in the environment and strategy. The learning approach of system dynamics is to use quantitative models and scenario runs, in order to explore, qualitatively, the behaviour of the system. Thus, it often suffices to show, with the help of a model, that barriers and optimal choices exist in order to improve discussions on strategic/tactic issues. Of practical importance are notions such as 'the optimal value for promotions shifts to the right (more promotions) when the total chain margin is increased'.

Creating a joint promotion budget

In this section, we calculate steady-state levels for the number of repeat (H) and occasional users (L), as given by equation 1d in Appendix 1. All other variables are computed according to the other algebraic functions.

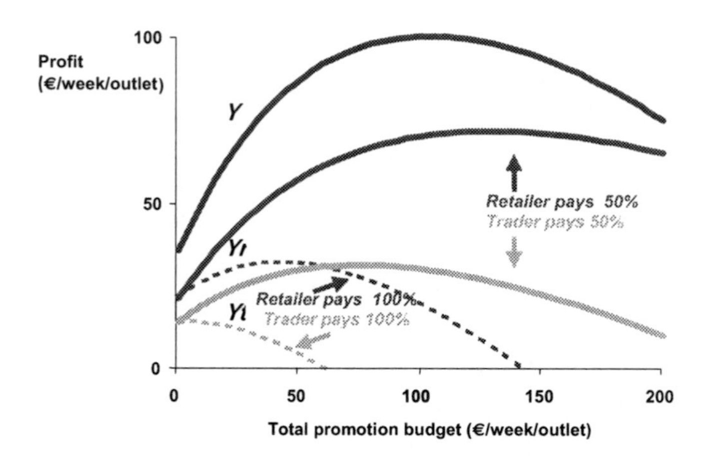

Figure 4. *Profits per chain player as function of the total promotion budget. Dashed lines indicate the profits – after the promotion cost – when the retailer or trader pays 100% of the promotion budget. Solid lines correspond to the profits when 50%/50% cost sharing of the promotion budget between retailer and trader is negotiated*

Figure 4 illustrates what the effect can be of cost sharing, in this case of the promotion activities. Promotions here denotes all activities that make consumers try the product (mathematically, move from the N to the L state). It may be giving away whole products to take home, have tasty pieces in the supermarket, advertising, etc. A simple free-rider problem exists, as promotions paid for by one player (e.g. the trader) automatically improve the profits of the other (the supermarket). In Figure 4, the dashed lines denote the profits, after deduction of the promotion budget of each player when they pay 100 % of the promotion cost themselves. Without promotions, profits for the retailer (Y_r) and trader (Y_t) are 14 and 21 €/week, respectively, per supermarket store.

The pay-off for the trader is unattractive; his profits would only decrease at every non-zero promotion budget. For the retailer, the profits do increase somewhat, until a promotion budget of 45 €/week/outlet, but the profit gain, from 21 to 32 €/week is too small to bother, given that a supermarket has many other products to attend to, and can afford to be critical in directing scarce resources.

Interestingly, the picture changes strongly when the trader and the retailer agree to share the cost of promotion, at 50%/50% for example. The reason is that the promotion cost per product item is offset by the total chain margin, and not by the margin of one player only. The maximum values of the solid curves for the pay-offs (profits) for the trader, the retailer and the combined profits (upper line, indicated with Y_c) lie around three times as high as is the case without promotion. The locations of these maxima differ, with the trader expected to prefer a promotion budget of around 80 €/week/outlet, whereas the retailer, if it would dare to fully optimize profits, would go as far as 130 €/week/outlet. However, the chain profit is maximal at 100 €/week/outlet.

Which player profits from ready-to-eat positioning?

Suppose now that a joint promotion budget has been agreed between the retailer and the producer of 65 € per week per outlet, shared 50%/50% between them. We now look at the positioning of exotic fruits as ready-to-eat (close to $T = 8$) instead of selling them at an on average unripe stage ($T = 5$). As Figure 3 showed, the dilemma is in avoiding product loss while satisfying consumer preferences. Therefore, we focus on the cost-sharing parameter f_w that determines which party accepts the risk of product loss. In the left-hand panel of Figure 5, the profit for each player is shown as a function of the average batch ripeness (T), both for the situation where the retailer pays for product loss ($f_w = 1$, dashed curves) and where the trader will take back all unsold product without cost to the retailer ($f_w = 0$, solid curves). The incentives to reposition (increase) the ripeness of sold fruit and to reduce the variation within each batch change strongly depending on this parameter, which identifies the bearer of product-loss risk (the retailer if $f_w = 1$, the trader if $f_w = 0$).

If the retailer pays for all product, whether sold to consumers or lost on its shelves ($f_w = 1$, dashed curves), as is commonly the case with fresh produce, the trader sees profits rise whether the fruits are eaten or lost, and the trade-off between limiting product loss and complying to consumer preferences applies to the retailer only. The retailer would find an optimal ripeness of around $T = 6.6$ days, even though the total chain profit has an optimum at $T = 7.2$ days. Alternatively, the trader could offer during negotiations not to charge for the volume that is lost on the retailer's shelves due to the ready-to-eat positioning, reflected here as $f_w = 0$ (solid curves). In that case the incentives change considerably, and both players should be expected to give ready-to-eat positioning a try with optimal values of T between 7 and 8 days (8 days being the middle of the acceptance region of consumers, their 'favourite ripeness'). This arrangement alone would not make the trader make much more profit, but it changes also the incentive of another profit-enhancing innovation that would otherwise not occur: reducing the variation of ripeness of fruits within a batch (parameter v). The right panel shows how profits increase as this variation is decreased (e.g. from $v = 5$ to $v = 2$, read from right to left).

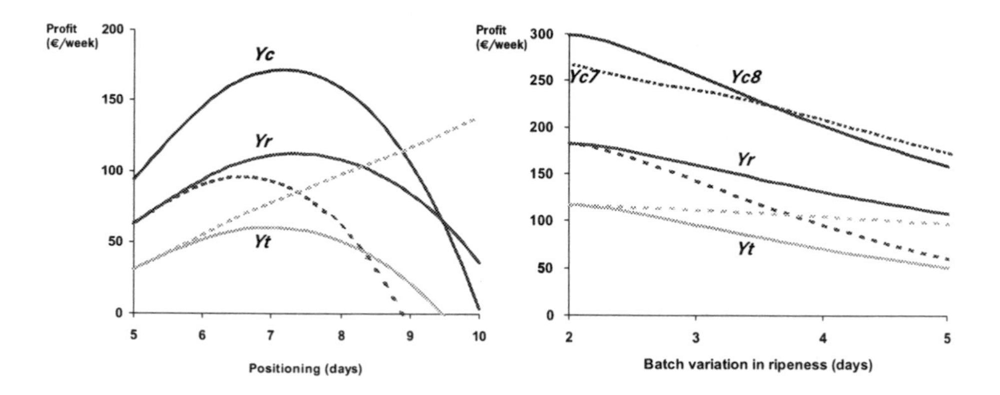

Figure 5. *Profits per chain player as function of the average ripeness (left) and variation in ripeness (right). Dashed curves indicate $f_w = 1$, solid curves $f_w = 0$. Left-hand panel is with variation $v = 5$ days, the right-hand panel is with average ripeness $T = 8$ days (except for dotted line indicated Yc7, which corresponds to $T = 7$)*

The dashed lines in the right hand panel represents the profits of the two players when the retailer bears the product-loss costs ($f_w = 1$), the solid lines denote $f_w = 0$. Again, a strong change in incentives occurs when the trader absorbs product-loss costs. The chain profit (sum of the profits of these two players) is denoted for both $T = 8$ (labelled *Yc8*) and for $T = 7$ (labelled *Yc7*), which allows to see that when the variation of ripeness is 'under control' ($v = 2$), the optimal positioning is again somewhat shifted towards the preference of consumers ($T = 8$), whereas with $v = 5$, at the far right of the right-hand graph, it would be better to stick to $T = 7$, as the variation is reducing profits through a large product-loss fraction.

The profits from optimizing product loss

Finally, we show how it is possible to determine the degree of product loss (W) that optimizes profit. In the left-hand panel of Figure 6, the chain profit *Yc* is shown as function of the product-loss fraction W, for four values of the variation in batch ripeness v. The right-hand panel first computes the positioning parameter T that results in the accepted product-loss fraction W on the x-axis. It results from solving equation 3 for T as a function of W. The profits are subsequently computed and plotted in the left-hand panel.

Although product loss can be seen as a consequence of product positioning on the ripeness dimension, as graphically shown in Figure 3, we may turn the argument in the opposite direction, in order to assess the optimal product positioning (shown to exist in Figure 5) as function of the product-loss fraction. We do this, because product loss may be easier to measure than consumer-preference parameters. The idea would be that the results in Figure 6 are more robust to uncertainty in estimates for parameter values than Figure 5. In Figure 6, the existence of an optimal level of product loss is shown (filled circles). At the same time, it demonstrates how this optimal level of product loss depends on variation control, and what the effect of reduced variation on profits is. As the variation is reduced from 5 to 2 days, profits increase and product loss decreases, even though the positioning (right-hand panel) is shifted towards riper fruits. Retailers can thus monitor product-loss fractions in relation to variation in ripeness, in order to adjust the ripening time of fruits in the supply chain, in order to optimize profits.

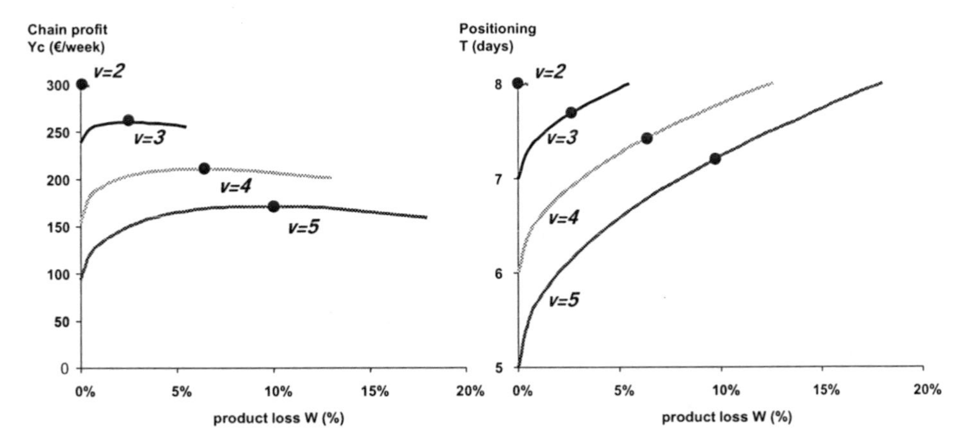

Figure 6. *Chain profit (Yc in Figure 5) as a function of product loss (W, given as percentage) is shown to have an optimum (•), dependent on the level of variation (v) of the quality attribute in the batches in the left-hand pane*

DISCUSSION

By combining product quality and its variation within batches of the product with the costs and benefits of different constituents in the production and supply chain of ready-to-eat exotic fruits and relating this to consumer behaviour, we make clear the peculiarity of highly perishable produce with biological variance. These types of chains occur mainly in the ready-to-eat food and ornamentals supply chains. From Figure 4 it is clear that vertical cooperation and risk/cost sharing can lead to simultaneous higher profits for different participants in these chains.

In this thought-experiment we use 'ripeness' to denote the apparently sole quality factor influencing consumer behaviour (aside from price, which is not varied in the model study for clarity's sake). However, 'ripeness' is itself a composite of several product attributes (texture, sweetness, smell, etc), and it is in fact meant to denote the Principal Component of all quality attributes of the product under study. In this way it is meant as a generic term for 'main axis of product quality' that can be influenced by the supply chain. The model shows that the most likely positioning of the product (e.g. here in terms of fruit ripeness) depends strongly on which trade partner (the retailer or the trader) takes responsibility for the loss of product. If the retailer pays for product loss, he will tend to position the product at an unriper stage, and failing to match the Liking function of the consumer, and therefore reducing the profit for both retailer and trader. If the trader takes the responsibility for product loss, both retailer and trader will tend to extend the positioning date, enhancing the Liking function and thereby increasing profit for both of them. This is strongly dependent on the amount of variation in ripeness in the batches of product. Figure 5 shows that at minimal levels of variability in quality ($v = 0$), profits for all parties are highest, but also that the profit difference between the cases where either the downstream or the upstream party takes responsibility for the product loss vanishes. This makes clear why these advantages of vertical cooperation in chains are not so clear in non-food chains, where variations in product quality within batches of the same product are usually negligible.

The model indicates that product loss due to over-ripeness in the retail should *still* be accepted, and even optimized. This approach is different from the classical approach in logistics where product loss *per se* is taken as the most critical factor and is minimized at all times. This is due to the fact that consumer behaviour is not included in these logistical calculations. However, it is clear from our model that a certain amount of product loss comes with an optimal positioning of ready-to-eat food in order to obtain the highest profit possible under the circumstances. Let us try to explain this: at present many exotic fruits are harvested at an unripe stage to allow long-term transport to distant outlets. If the fruits become ripe too early they usually start producing volatile plant hormones such as ethene (ethylene). These gasses tend to trigger autocatalytic ripening processes in nearby fruits. As such, whole cargoes of fruits can be destroyed en route. As a consequence the fruits generally do not reach the proper ripening stage even when the consumer buys them and the consumer will be disappointed in the taste of the fruit. If the vertical participants in the chain were to cooperate on the level of fruit ripeness and variation in this ripeness factor, the optimal profit for all partners in the chain could be elevated and the consumer would be more satisfied with the product. The model shows the advantage all participants in the chain can obtain from this cooperation and as such can be used as a tool to support negotiations between potential chain partners to the benefit of all chain participants. The model shows that classical supply-chain management approaches, focused on limiting product loss, yield a sub-optimal profit level. The modelling exercise makes it clearer how making consumer preferences the leading factor behind product revenues, further increases total chain profits. It is therefore meant as a main step towards consumer-driven product development.

Obviously, further research is necessary, especially in the area of calibration and validation. A large agenda for further research has been drawn up, as the current model can function as a backbone of tying the different disciplines together. Addressing the more specific dynamics of fruit ripening and its effect on keeping quality and product loss is our next step. With mango for instance, it is suspected that mangoes harvested before a critical ripeness, will never ripen properly while in the supply chain. The fruit will deteriorate earlier. How these effects work out for various fruits and are affected by the variance in ripeness within batches is critical (see Tijskens and Polderdijk (1996) and Tijskens et al. (2003)). When these issues have been addressed to some degree of detail, heuristics may be derived that can guide practical implementation of fresh exotic-fruit chains.

REFERENCES

Schepers, H.E., Van Henten, E.J., Bontsema, J., et al. 2004. Tactics of quality management and promotions: winning consumers for fresh exotic produce. *In:* Bremmers, H.J., Omta, S.W.F., Trienekens, J.H., et al. eds. *Dynamics in chains and networks: proceedings of the sixth international conference on chain and network management in agribusiness and the food industry (Ede, 27-28 May 2004)*. Wageningen Academic Press, Wageningen, 568-582.

Sterman, J.D., 2000. *Business dynamics: systems thinking and modelling for a complex world.* McGraw-Hill, New York.

Tijskens, L.M.M., Konopacki, P. and Simcic, M., 2003. Biological variance, burden or benefit? *Postharvest Biology and Technology,* 27 (1), 15-25.

Tijskens, L.M.M. and Polderdijk, J.J., 1996. A generic model for keeping quality of vegetable produce during storage and distribution. *Agricultural Systems,* 51 (4), 431-452.

APPENDIX 1: MODEL EQUATIONS

The dynamics of the number of repeat users (H). occasional users (L) and non-users (N) is described by the following differential equations:

$$\frac{dH}{dt} = QE_Q L - H/\tau_H \tag{1a}$$

$$\frac{dL}{dt} = -(QE_Q L - H/\tau_H) + (UE_U + r_L)N - L/\tau_L \tag{1b}$$

$$N(t) = Z - L(t) - H(t) \tag{1c}$$

where $Z = N + L + H$ is the total number of consumers. Product quality Q and promotions budget U (see Table 1 for units) are the management levers that determine the size of the consumer segments. E_U is the promotion effectiveness, and r_L is the spontaneous rate of trying the product. Solving equation (1a-c) gives the steady-state solution

$$\frac{L}{Z} = \frac{1}{1 + \left[(r_L + UE_U)\tau_L\right]^{-1} + QE_Q\tau_H} \tag{1d}$$

from which the steady-state values for H and N are found easily. The product quality Q is the integral from the minimum acceptance level (a) to the maximally accepted perceived ripeness (m) of the variation distribution of supplied fruit ripeness represented by the piecewise 'tent' function, the horizontally hatched area in Figure 3a:

$$\begin{cases} Q = 1/2\left(\dfrac{T + v + a}{v}\right)^2 - 1/2\left(\dfrac{T + v - m}{v}\right)^2, a > T \\ Q = 1 - 1/2\left(\dfrac{T + v - m}{v}\right)^2 - 1/2\left(\dfrac{v + a - T}{v}\right)^2, a < T \end{cases} \tag{2}$$

Finally, W is the product-loss percentage due to unacceptable fruit, v being the variation within each batch of fruit with respect to firmness (in days), and m is the maximum firmness limit (too soft) that consumers accept. It determines the point where product loss starts. W is the vertically hatched area in Figure 3 (left-hand panel), and its expression is

$$W = 1/2\left(\frac{T + v - m}{v}\right)^2 \tag{3}$$

for $T+v>m$ and otherwise zero.

The outputs of the model are the profit levels of the chain players, Y_r for the retailer, Y_t for the trader and Y_p for the producer. The profit of the retailer is calculated as the product of sold volume and gross margin, minus share of promotion cost. The volume V bought by consumers equals

$$V = (D_H H + D_L L)d^e \tag{4}$$

where D_H and D_L are the intended usage frequencies for fruit for H and L, respectively (for values and dimensions see Table 1 below), d (dimensionless) is the relative price position taken by the retailer, and e is the price elasticity of consumers (all segment being equal in this respect). The profit of the retailer is found after multiplication by the retailer gross margin (in square brackets) and subtracting its share of promotion cost (which is here the only product-related fixed cost considered):

$$Y_r = V\left[dP_{cr} - \frac{P_{rt}}{1 - f_w W} - \frac{C_{hp} + TC_{hd}}{1 - W}\right] - (1 - u_t)U \tag{5}$$

where P_{cr} is the consumer reference (expected) price, P_{rt} is the 'transfer price' or purchase price between the retailer and the trader, C_{hp} is the fixed handling cost per piece of fruit by the retailer, C_{hd} is the handling cost per piece per day that the retailer has the fruit in storage, before display on store shelves. The fraction of the product-loss volume that the retailer has to pay for to the trader, f_w, thus only affects the purchased volume, not the cost of the handled volume. Finally, u_t is the share (percentage) of promotion cost that the trader pays. The profit of the trader follows as

$$Y_t = \frac{V}{1 - f_w W} P_{rt} - \frac{V}{1 - W} P_{tp} - u_t U \tag{6}$$

where the volume sold to the retailer is adjusted for the product-loss fraction f_w, whereas all volume, whether bought by consumer or wasted at the retailer, has been paid to the producer/exporter. P_{tp} is the transfer price between exporter and trader. The profit of the producer has the expression:

$$Y_p = \frac{V}{1 - W} [P_{tp} - C_p] \tag{7}$$

with C_p being the unit cost to the exporter. Adding up the profits of the trader and retailer, the 'marketing chain', the chain profit Y_c becomes

$$Y_c = V \left[dP_{cr} - \frac{C_{hp} + TC_{hd} - P_{tp}}{1 - W} \right] - U \tag{8}$$

and is, as expected, independent of the cost-sharing parameters f_w, u_t and the transfer price P_{rt}.

THE VALUE OF INFORMATION IN AGRI-FOOD CHAINS

CHAPTER 10

INFORMATION MANAGEMENT IN AGRI-FOOD CHAINS

G. SCHIEFER

University of Bonn, Meckenheimer Allee 174, D-53115 Bonn, Germany.
E-mail: schiefer@uni-bonn.de

Abstract. The establishment and management of information infrastructures in chains and beyond is a prerequisite for the implementation of the emerging comprehensive requirements on tracking, tracing and quality assurance in agriculture and the food sector. They support the guarantee of food safety and the focus on consumers' quality needs. The challenge for the sector is the agreement on, and the implementation of, appropriate information infrastructures. The paper discusses the issue by extending the classical enterprise information hierarchy by two additional information layers that cross the enterprise boundaries and form a sector-wide information network.
Keywords: information management; chains; tracking and tracing; quality

PROBLEM SCENARIO

Traditionally, information management in enterprises builds on a number of information layers that correspond with the different levels of business management and decision support. They reach from transaction information at the lowest level to executive information at the highest level (e.g., Turban et al. 1999). These traditional layers are presently being complemented by two additional layers at the lower, transaction, level that incorporate information for tracking and tracing and for quality assurance and improvement activities (Figure 1).

These new layers differ from traditional enterprise information layers by their focus. Their focus is not the individual enterprise but the vertical chain of production and trade. They are linked to the flow of goods and connect, in principle, the different stages of production and trade with each other and the consumer. The layers were initiated by requirements from legislation and markets for:

a. tracking and tracing capabilities (see, e.g., EU regulation 178/2002); and
b. increased consideration of consumer needs and expectations regarding the quality of products and production processes.

C.J.M. Ondersteijn et al. (eds.), Quantifying the agri-food supply chain, 137-146.

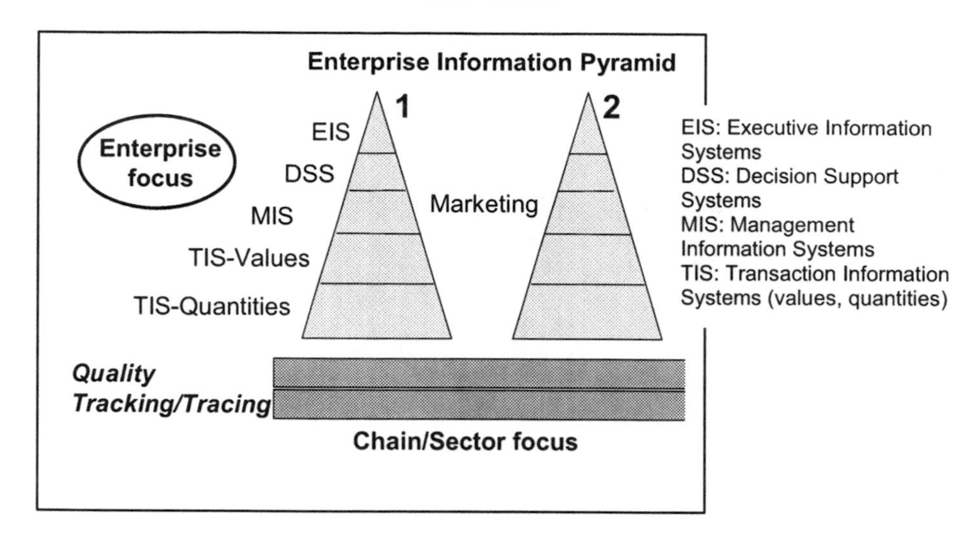

Figure 1. *Information layers with enterprise (1,2) and chain/sector focus*

Both, tracking and tracing capabilities as well as the fulfilment of quality expectations, depend on activities throughout the supply chain and, as a consequence, on communication between the various stages of the chain and the establishment of a chain or sector information infrastructure.

The implementation of tracking and tracing capabilities as well as the fulfilment of quality expectations involves chain and sector efforts and agreements on who does what, when, where and how. This complex scenario involves efforts on:

a. negotiation, mediation, engagement and investment;
b. the adaptation of process organizations to match the requirements on tracking, tracing and quality assurance;
c. the organization, management and operation of the information exchange including of necessary interfaces, data networks, data bases, data-processing activities, and data utilization through business intelligence products.

These efforts correspond with costs. The heterogeneity of the sector with its different but interlinked production lines and the diversity of enterprises place high costs on the negotiation, mediation, engagement and investment efforts. Furthermore, some parts of the sector, especially the commodity sector with its classical bulk products, might be forced to engage in major process reorganizations to meet expectations on its tracking/tracing capability and to keep costs in case of food quality failures under control.

Legal aspects put pressure on the sector to initiate the efforts. However, individual enterprises might consider the initiating costs too high when compared with potential individual market benefits from an improved tracking/tracing and quality assurance capability. They might disregard new opportunities for utilizing the infrastructure for improvements in quality assurance, chain coordination and chain management. As a consequence the sector might be forced to enter a step-by-step-development path that builds on individual development clusters of innovator enterprises instead of identifying and implementing a comprehensive best solution. This could reduce the initiating costs but increase costs for the organization, management and operation of the information exchange.

In this paper, the main focus is on the organization, management and operation of the information exchange, summarized in the following as information management. In addition, it will take up some aspects of process organization that are closely interlinked with information-management issues and extend the discussion towards the potential benefits of the new information infrastructure.

The paper's main goal is to identify the need for information infrastructures that evolve from sector developments, to evaluate the managerial implications of potential and actual development alternatives, and to link the infrastructure developments with potential benefits beyond the actual development drivers.

THE FOCUS: INFORMATION INFRASTRUCTURE EXTENSION

The establishment of information infrastructures for enterprise communication is not a unique or new scenario. E-commerce and the digital exchange of trade documents in business transactions have been the focus of much attention in the agri-food business community since many years (Schiefer et al. 2003). However, the establishment of information infrastructures for tracking/tracing capabilities and quality assurance are different and much more complex tasks.

While agreements on the communication of trade documents primarily depend on agreements on technical specifications (see, e.g., the agreements on the EAN codes and the EDIFACT document exchange format (Kuhlmann 2003), communication on the new information layers is closely related to business policies in a competitive business environment. A sector-encompassing general agreement is restricted to the lowest level of legal requirements. Any communication agreements beyond this level are subject to specific business interests and might limit themselves to clusters of enterprises with common trading interests. In a network environment, individual enterprises might be members of different clusters, resulting in a patchwork of interrelated and overlapping communication clusters (Figure 2).

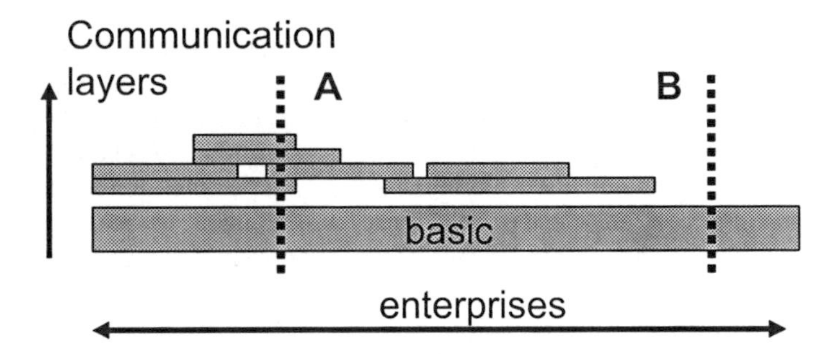

Figure 2. *Agreed communication clusters and examples for enterprise participation in five (enterprise A) and one (enterprise B) of the clusters*

The feasibility and 'value' of the information infrastructure depend on participation of each individual enterprise within a group of similar interests. In a net chain environment with open and changing business relationships (Lazzarini et al. 2001), the level of legal tracking and tracing communication requirements may involve most of the enterprises within a certain food sector on a national or even global scale. The dependency between enterprises makes the value of the infrastructure dependent on the weakest link. This forces the sector into the establishment and management of a generally accepted and implemented basic communication layer that leaves room and provides the format for higher levels of agreements between participating subgroups.

This structural model might be the basis for a general sector solution or, alternatively, for independent infrastructure clusters that might be implemented independently of each other by different groups. From a sector point of view, the first alternative requires a higher degree of agreement throughout the sector but is characterized by simplicity in system organization and management. The second alternative allows individual initiatives to develop independently. This reduces the initial need for sector-wide agreements but adds coordination complexity in system organization and management. However, whatever the development path, the principal problems in the design, establishment and management of the information infrastructure are the same.

INFORMATION ORGANIZATION

Organization level: tracking and tracing

The information for tracking and tracing involves an enterprise and a chain dimension. The information is linked to the flow of goods. Within agri-food enterprises, traditional ERP (Enterprise Resource Planning) solutions do not support the monitoring of individual product items or individual batches in commodities. The integration of this aspect into ERP solutions is a software development issue that does not require any chain- or sector-wide agreement initiatives.

The major challenge is the monitoring of individual products or batches on their path through the vertical supply chain of trading partners. In trading environments with a well-defined and limited number of potential trading partners, as is the case with closed supply chains, the establishment of an appropriate information infrastructure could be built on agreements by the trading partner group (see Figure 3).

However, in a net chain environment with continuously changing trading partners, the chain communication model (Figure 3) represents agreements within one of the communication clusters of Figure 2 that need to build on a basic communication layer that extends the chain approach to the whole trading environment.

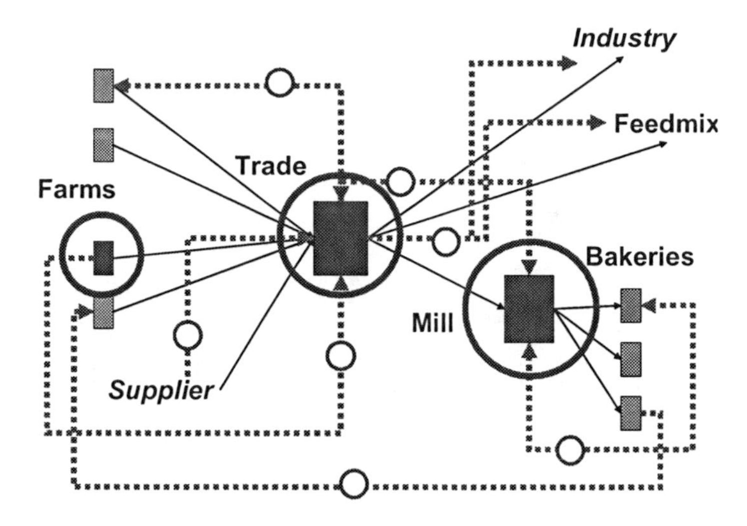

Figure 3. T&T information infrastructure for a chain-based tracking and tracing capability

Organization level: quality assurance

The quality information layer adds content to the tracking and tracing capability. Information for the support of quality assurance of products towards the consumer as the final customer builds on enterprise-internal requirements, the requirements of the direct customer, and the requirements of the consumer as the final customer.

The diversity of interests could generate an almost unlimited number of possible requirement sets. However, the sector builds on a limited number of quality systems that incorporate certain sets of quality requirements (see, e.g., Krieger and Schiefer 2004). Some of these quality systems are widely accepted in the sector and incorporate an invaluable degree of agreement regarding the relevance of quality characteristics. Furthermore, some of these systems build on a chain view and cover the different stages of the supply chain. Examples are the IKB system in The Netherlands and the Q&S system in Germany. Other systems like the IFS system of retail focus on retail's immediate suppliers but influence, indirectly, the whole supply chain of the suppliers.

These quality systems could serve as a basic reference for different levels of quality communication within the quality information layer. First initiatives towards this end are under way. They include the organization of databases with enterprise information of groups of enterprises that participate in certain quality systems. The establishment of different levels of quality communication would separate the

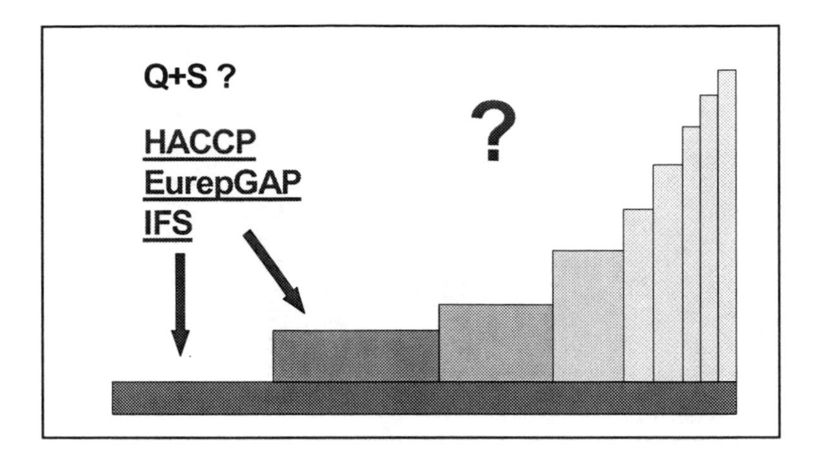

Figure 4. Production segments with different levels of quality guarantees

sector's food production into different segments with different quality guarantees (Figure 4).

To structure communication in a well-organized information infrastructure, one could take advantage of the fact that the quality aspects in quality systems correspond with four different layers of quality focus. The quality aspects in quality systems may focus on the quality of products, the quality of process organization, the quality of process management or the quality of enterprise management (Figure 5). This approach supports the integration of different quality systems.

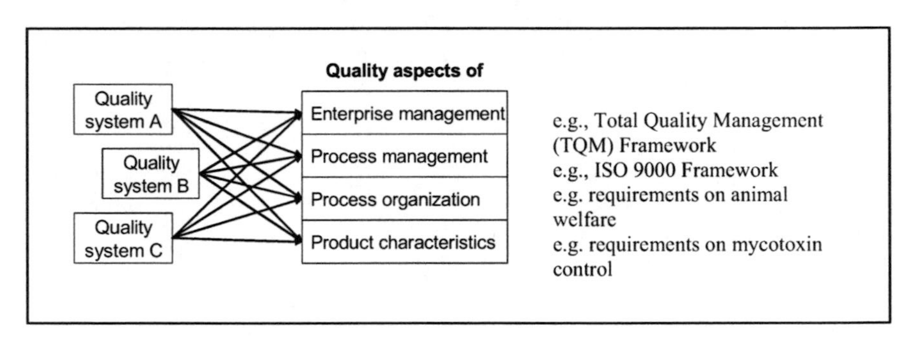

Figure 5: Layers of quality focus (see Schiefer and Rickert 2004)

INFORMATION INFRASTRUCTURE

The principal alternatives for sector-wide information infrastructures focus on two different dimensions.

The information may be communicated between enterprises through a common data network that is linked with enterprises' internal information systems (see Figure 6). Alternatively, the information may be communicated between enterprises directly as shown in Figure 3. These approaches mirror classical network approaches as, e.g., bus or ring network topologies (Turban et al. 1999).

The second dimension concerns the initiation of the communication. Information might be communicated on demand (trigger system) or, alternatively, the information might be communicated any time according to specified communication rules irrespective of actual needs.

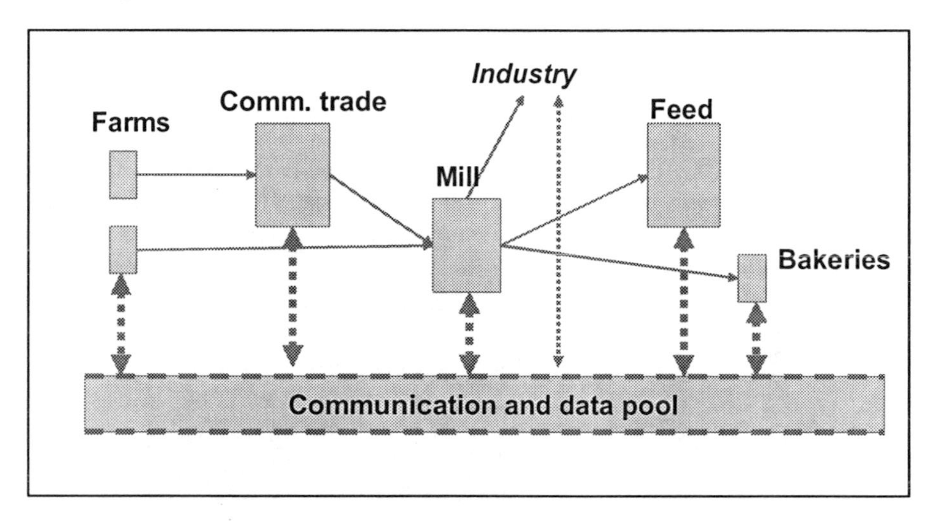

Figure 6*. Sector-wide communication and data pool (example: grain chain)*

As a consequence, the information infrastructure could build on any of four organizational alternatives (Table 1).

Table 1*. Information infrastructure alternatives*

	I: data pool communication	II: enterprise linkage
A: Trigger system	A-I	A-II
B: Rule system	B-I	B-II

The different information layers could follow different organizational approaches. Actual but not yet published developments focus on:

a. the organizational approach A-I for tracking and tracing purposes; and

b. the approach B-I for quality assurance communication.

The system alternatives A-II and B-II are reported from small groups of closely cooperating enterprises that directly link their ERP systems for data communication.

However, there is an additional alternative of communication that avoids the communication of data but communicates assurances that certain information is true. If enterprises are assured that their suppliers fulfil the requirements of a certain quality system, information linked to the requirements do not have to be communicated; the assurance (e.g. in terms of a certificate) is sufficient. As information infrastructures for quality assurance are not yet established sufficiently, this last approach is being implemented with a number of quality systems. An example is the Q&S system (Nienhoff 2004) (Figure 7).

ADDED VALUE OF NEW INFORMATION INFRASTRUCTURE

In judgments regarding the costs and benefits of the newly developing information infrastructure one needs to keep in mind that the existence of such an infrastructure would greatly facilitate the implementation of a variety of information services for business support.

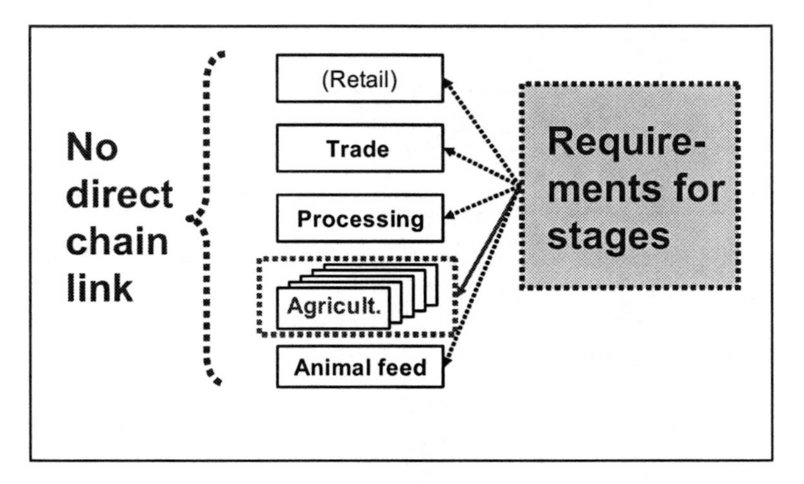

Figure 7. Communication through certification regarding the fulfilment of requirements (see Schiefer 2004)

Typical and basic examples include:

a. the organization of chain focused consulting services;
b. the communication of quality guarantees to consumers;
c. the establishment of a chain inventory management activity.

The new information layers provide the information basis for chain-focused consulting (extension) services regarding improvements in quality. As an example, the quality information layer could incorporate production information from farms as well as information from processing regarding the suitability of farm deliveries etc. This information combined and related to external information from other sources might better enable consulting services to advice farms on changes in production processes.

The communication of quality guarantees to consumers beyond what is available on labels has gained interest with food enterprises. As a basic example, the ability to offer consumers a chance to check for themselves the origin of their products (e.g., farms) or a product's supply-chain path might become a competitive business advantage (see, e.g., Boeve 1999; Schiefer et al. 1999).

The establishment of a chain inventory management activity is a first step towards more sophisticated chain management initiatives that might utilize the new information infrastructure. Chain inventory management builds on the exchange of information on the availability of inventories at the various stages of the supply chain. As an example, a grain-processing enterprise might greatly benefit in its own production and sales' activities from information on grain inventories and their quality in its supplier farms.

All examples involve certain aspects of chain management for improved chain efficiency that depend on the availability of an information infrastructure. The quality interest, the chain management aspect and the legal requirements on the tracking and tracing capability of the food chain together provide the argument for the establishment of a sector-wide information infrastructure. These benefits combined are the long-term matching part for the costs of a sector-wide information infrastructure.

CONCLUSION

The need for new management approaches in food supply chains, especially regarding food safety guarantees and quality assurance activities, requires new initiatives in information management. At the core of interest is the need for new information layers that utilize enterprise information but focus on the communication between chains for quality assurance towards the consumer as the final customer and for improvements in risk management and tracking or tracing capability in case of problems in food safety or quality.

As some of these aspects have become legal requirements, the sector is forced to act. However, as the balance of costs and benefits for individual enterprises regarding general sector solutions might be low or even negative, it is suspected that solutions will have to build on a network of enterprise clusters of limited size that are easier to coordinate for utilizing some of the potential benefits of the new information layers that reach beyond basic legal requirements and might involve improvements in food quality, in tracking and tracing, and in chain efficiency.

Several projects not yet published in literature are under way. They might serve as a basis for a sector-wide network of clusters, a semi-optimal but feasible solution for meeting the sector's information management needs in the future.

REFERENCES

Boeve, A.D., 1999. Integrated veal information IVI. *In:* Schiefer, G., Helbig, R. and Rickert, U. eds. *Perspectives of modern information and communication systems in agriculture, food production and environmental control: proceedings of the second European conference of the European Federation for Information Technology in Agriculture, Food and the Environment, September 27-30, 1999, Bonn, Germany / EFITA*. University of Bonn/ILB, Bonn, 835-844.

Krieger, S. and Schiefer, G., 2004. Quality management schemes in Europe and beyond. *In:* Schiefer, G. and Rickert, U. eds. *Quality assurance, risk management and environmental control in agriculture and food supply networks: proceedings of the 82nd European seminar of the European Association of Agricultural Economists (EAAE), May 14-16, 2003, Bonn, Germany*. University of Bonn/ILB, Bonn, 35-50.

Kuhlmann, F., 2003. EAN enabling technologies for the meat supply chain. *In:* Schiefer, G., Helbig, R. and Rickert, U. eds. *E-commerce and electronic markets in agribusiness and supply chains: proceedings of the 75th seminar of the European Association of Agricultural Economists, EAAE, February 14-16, 2001, Bonn, Germany*. University of Bonn/ILB, Bonn, 219-226.

Lazzarini, S.G., Chaddad, F.R. and Cook, M.L., 2001. Integrating supply chain and network analyses: the study of netchains. *Journal on Chain and Network Science*, 1 (1), 7-22.

Nienhoff, H.J., 2004. QS quality and safety: a netchain quality management approach. *In:* Schiefer, G. and Rickert, U. eds. *Quality assurance, risk management and environmental control in agriculture and food supply networks: proceedings of the 82nd European seminar of the European Association of Agricultural Economists (EAAE), May 14-16, 2003, Bonn, Germany*. University of Bonn/ILB, Bonn, 627-630.

Schiefer, G., 2004. From enterprise activity 'Quality Management' to sector initiative 'Quality Assurance': development, situation and perspectives. *In:* Schiefer, G. and Rickert, U. eds. *Quality assurance, risk management and environmental control in agriculture and food supply networks: proceedings of the 82nd European seminar of the European Association of Agricultural Economists (EAAE), May 14-16, 2003, Bonn, Germany*. University of Bonn/ILB, Bonn, 3-22.

Schiefer, G., Helbig, R. and Rickert, U. (eds.), 1999. *Perspectives of modern information and communication systems in agriculture, food production and environmental control: proceedings of the second European conference of the European Federation for Information Technology in Agriculture, Food and the Environment, September 27-30, 1999, Bonn, Germany / EFITA*. University of Bonn/ILB, Bonn.

Schiefer, G., Helbig, R. and Rickert, U. (eds.), 2003. *E-commerce and electronic markets in agribusiness and supply chains: proceedings of the 75th seminar of the European Association of Agricultural Economists, EAAE, February 14-16, 2001, Bonn, Germany*. 3rd edn. University of Bonn/ILB, Bonn.

Schiefer, G. and Rickert, U. (eds.), 2004. *Quality assurance, risk management and environmental control in agriculture and food supply networks: proceedings of the 82nd European seminar of the European Association of Agricultural Economists (EAAE), May 14-16, 2003, Bonn, Germany*. University of Bonn/ILB, Bonn.

Turban, E., McLean, E. and Wetherbe, J., 1999. *Information technology for management*. Wiley, New York.

CHAPTER 11

THE INSURABILITY OF PRODUCT RECALL IN FOOD SUPPLY CHAINS

MIRANDA P.M. MEUWISSEN[#], NATASHA I. VALEEVA[#,##], ANNET G.J. VELTHUIS[##] AND RUUD B.M. HUIRNE[#,###]

[#]*Institute for Risk Management in Agriculture, Wageningen University and Research Centre, Hollandseweg 1, 6706 KN Wageningen, The Netherlands. E-mail: miranda.meuwissen@wur.nl*
[##]*Business Economics Group, Wageningen University, Wageningen, The Netherlands*
[###]*Animal Sciences Group, Wageningen University and Research Centre, Wageningen, The Netherlands*

Abstract. Insurers face growing difficulties with insuring food-related risks among others due to an increasing number of product recalls and an increasing amount of claims being pushed back into the chain. This paper focuses on the risk of product recall in dairy supply chains. The paper aims at providing insurers with useful tools for insurance design and claim handling. More specifically, the objectives of this paper are (1) to define product recall, aimed at recognizing recall perils and losses; (2) to identify important precautionary action points and related control measures, for underwriting and class-rating purposes; (3) to develop a risk assessment framework, as a tool for calculating premium levels; and (4) to evaluate third-party verifiability of due diligence, aimed at identifying eligibility for insurance payments. Precautionary action points are prioritized with adaptive conjoint analysis. In the risk assessment framework, case studies are used to quantify the size of losses. Additionally, throughout the paper, expert consultation has been an important source of information. Results show that perils and losses of product recall need to be strictly defined, preferably on a case-by-case basis. Also, case studies show that recall losses easily cumulate as losses are yet between Euro 210,000 and Euro 2,300,000 for only a limited number of recall expenses and contaminated products. Furthermore, in relation to the third-party verifiability of due diligence, difficulties are encountered at the farm level. We conclude that, if risks are properly defined and insurance schemes incorporate adequate due diligence and disclosure incentives for all chain participants, product recall remains an insurable type of risk, even if the number of recalls will further increase.
Keywords: precautionary action points; risk assessment; due diligence; dairy supply chains; compound feed; insurance

INTRODUCTION

Insuring liability and recall risks in food supply chains is getting increasingly complex. On the one hand, risk prevention gets lots of attention, therewith reducing

C.J.M. Ondersteijn et al. (eds.), Quantifying the agri-food supply chain, 147-159.

the risk of food safety crises and related liability claims (Segerson 1999; Henson and Hooker 2001; Valeeva et al. 2005). On the other hand, the number of recalls is increasing (Teratanavat and Hooker 2004) and traceability systems allow claims to be pushed back into the chain (Meuwissen et al. 2003). At the same time, third-party verifiability of due diligence is getting increasingly important, both for counterattacking liability claims and for proving the unintentional character of a (recall) loss. Also, with the 2005 implementation of the General Food Law (EC/178/2002) adequate performance with respect to traceability and recalls is no longer facultative but legally required.

Due to such changes in the risk environment of food supply chains, insurers face growing difficulties in designing adequate insurance schemes and in calculating proper premium levels. As a consequence, they may opt for higher risk loadings or an increasing number of perils and losses excluded from cover, therewith reducing the insurability of food-related liability and recall risks. A reduced availability of insurance cover is generally not considered beneficial (Arrow 1996). Skees et al. (2001) specifically address the positive incentives of recall insurance for improving the level of food safety.

Of the various food-related risks, this paper focuses on the risk of product recall. Recall risks are fairly straightforward and work well as an illustration for other food-related risks such as liability risks. We aim to provide insurers with a number of useful tools for debating the future feasibility of insuring product recall. More specifically, the objectives of this paper are:

(1) To define product recall, aimed at recognizing perils and losses.
(2) To identify important precautionary action points and related control measures, for underwriting, class rating and specifying proper rules of behaviour.
(3) To assess the risk of product recall, as a tool for calculating premium levels.
(4) To evaluate third-party verifiability of due diligence, aimed at identifying eligibility for insurance payments.

The objectives (1), (2) and (3) are relevant from an *insurance designing* point of view, while objective (4) is a crucial aspect of *claim handling*.

These issues are studied for dairy supply chains in The Netherlands and, more specifically, for the supply chain of fluid pasteurized milk. This chain is characterized by a few large supplying and processing industries and many small dairy farms (CBS and LEI 2005). Throughout the chain a lot of attention is paid to quality control and assurance (Valeeva et al. 2005) and Dutch consumers believe pasteurized milk to be a very safe product (Novoselova et al. 2002). Recently, only a few recalls have taken place, viz., with respect to penicillin (2001 and 2002) and hydrogen peroxide (2002).

MATERIALS AND METHODS

In relation to the first objective, i.e., a checklist for recognizing potential perils leading to a product recall and their related losses, a review was made of relevant internet pages (European Food Safety Authority, Dutch Food Safety Authority), insurance programmes (also through internet) and the General Food Law (EC/178/2002).

With regard to the second objective, a list of important precautionary action points and related control measures for underwriting purposes, we studied three chain participants, viz., feed companies, dairy farms and dairy industries, and two food safety perils: chemical hazards and microbiological hazards. More specifically, the chemical hazards include antibiotics and dioxin and the microbiological hazards refer to *Salmonella, E. coli, S. aureus* and *M. paratuberculosis*. The specification was done after literature research (Cullor 1995a; Collins 1997; Cullor 1997; Gould et al. 2000; Mathews et al. 2001) and consultation with representatives from the dairy industry, research organizations and regulatory authorities. Relevant precautionary action points and related control measures were selected based on a review of the scientific literature (among others Cullor 1995b; Sischo et al. 1997; Veling et al. 2002), current regulations, and individual consultations with experts from the various chain participants considered. In order to prioritize the identified action points, two workshops were organized in October 2002. A total of 22 respondents participated in these workshops. These were four experts from feed companies, thirteen of dairy farms and five of dairy industries. During the workshops respondents had to fill in computerized questionnaires, following the adaptive conjoint analysis (ACA) technique. A more extensive description of these workshops and materials and methods used is given by Valeeva et al. (2005).

In relation to the third objective, i.e., a framework for risk assessment and rating, we focus on the loss part of product recall in dairy supply chains; the *probability* of these losses occurring is not considered. Losses are assessed through: (1) estimating the size of relevant batches in dairy supply chains, and related processing times; (2) calculating the lost value of destroyed products and the costs of handling and notifications; and (3) assessing four case studies varying in point of contamination and allocation. Besides the three chain participants considered previously, this part of the study includes retail and consumers as well. Batch sizes, processing times and recall expenses were verified by farm and industry experts.

For the fourth objective, assessing third-party verifiability of due diligence as an instrument for evaluating the eligibility for insurance payments, we focused on the top-five action points resulting from the ACA workshops. This implies a total number of 30 precautionary action points and their related control measures. The third-party verifiability of the control measures was assessed on a three-point scale, i.e.,'fully verifiable', 'partly verifiable' or 'not verifiable at all', and checked with a lead auditor of an accredited certification body regularly auditing food supply chains. At this stage two mid-points, i.e., 'fully/partly verifiable' and 'not/partly verifiable' were deemed necessary to adequately reflect third-party verifiability circumstances at the farm level.

IDENTIFYING PERILS AND LOSSES

Insurers generally insure named perils and losses or, in case of 'all-risks insurance' there are usually a number of perils and losses specifically excluded. In relation to the perils leading to a recall, existing insurance schemes often cover 'product recall due to contamination'. This is clearly a food safety issue. However, the 'non-compliance with food safety requirements', as stated in the General Food Law (EC/178/2002, article 19) obviously includes more aspects than only contamination: "if a food business operator considers or has reason to believe that a food which it has imported, produced, processed, manufactured or distributed is *not in compliance with the food safety requirements*, it shall immediately initiate producers to withdraw the food in question from the market where the food has left the immediate control of that initial food business operator and inform the competent authorities thereof (..)".

Besides hazards leading to 'public health at risk', Table 1 lists a number of other food safety aspects that can lead to product recall, such as 'not fit for human consumption', referring to, i.e., spoilage of products and 'faulty claim on label'. An example of the latter is 'fit for diabetics', while due to some mistake the product is actually not fit for this group of consumers. In addition, besides food safety reasons, there can be other motivations for product recall as well, such as non-compliance with aspects of quality or image.

Table 1. *Perils and losses of product recall*

Perils	Scope of losses	Types of losses
Non-compliance with respect to: ▪ *food-safety* requirements, i.e.: - public health at risk - not fit for human consumption - non-compliance with legislation - faulty claim on label ▪ *Quality* issues ▪ Aspects of *image*	▪ *Customer* level: - non-conforming product(s) or batch(es) - suspected product(s) or batch(es) ▪ *Further along the chain*: - non-conforming product(s) or batch(es) - suspected product(s) or batch(es)	▪ *Producer* recall expenses, such as: - decreased value of products - product handling - notifications to customers or end users - relocation of the product - business interruption - rehabilitation expenses ▪ *Customer* recall expenses, such as: - business interruption - increased cost of production - empty shelves ▪ Recall expenses and liability claims *further along the chain*

In relation to the scope of losses, it is obvious to recall *non-conforming* products at customer and consumer level. However, also *suspected* products or batches are generally recalled or destroyed. For any insurance scheme covering losses from 'suspected' batches, it is crucial to define carefully what is meant by this term. For instance, if a dairy farm uses various sorts of compound feed and one appears to be contaminated, does this always mean that all raw milk needs to be destroyed? Also, if a dairy industry recalls a specific batch of consumer milk, does retail always agree with recalling only this specific 'best before' date or do they also return other batches or even other products from the company? And if so, does the insurer provide cover for this? With respect to the types of losses, Table 1 lists a number of recall expenses, both for producers and customers. For instance, producer recall expenses include the decreased value of the recalled products (values not necessarily reduce to zero as other usages may be possible), costs of handling and notifications, costs of relocating the product, and losses of business interruption and brand rehabilitation. At the customer level there may also be losses of business interruption (e.g., dairy farmers not being able to deliver milk because of some feed contamination), increased cost of production (such as dairy farms facing higher culling rates due to contaminated feed) or retailers being confronted with empty shelves. Similar costs may occur further along the chain, possibly leading to liability claims.

Clearly, although product recall seems to be a straightforward type of risk, it is crucial to define strictly the perils and (scope and types of) losses covered. The wide variety of perils and losses may induce insurers not to generalize 'product recall' across chain participants and supply chains but to define these issues on a case-by-case basis.

IMPORTANT PRECAUTIONARY ACTION POINTS FOR UNDERWRITING

From the review and expert consultation, we identified 82 precautionary action points for the chemical and microbiological hazards considered. More specifically, from these action points, 6 were relevant for chemical hazards, 41 for microbiological hazards and 35 for both. For chemical hazards, more than 50% of the identified action points are at the feed level, whereas 32% and 17% relate to the dairy farm and dairy industry, respectively. For microbiological hazards, the number of action points per chain participant is more equally distributed, but with a focus (40%) on farm level.

For compound feed production, precautionary action points and related control measures refer to purchase, transport and storage of compound feed ingredients and the identification and traceability of both feed and its ingredients. Also there are action points relating to the design of production facilities at the feed plant and the production practices and hygiene conditions for compound feed production and transport. In relation to the dairy farm, precautionary action points include a wide variety of aspects, viz., the purchase and production of feed, the grazing of pastures, cattle movement and its traceability, herd health and treatment, dairy-cattle housing, calving and feeding of calves, water management, and general hygiene conditions at the farm. For dairy industries precautionary action points identified refer to, i.e., transport of raw milk to the processing factory, the design of production facilities at the dairy plant, production practices and hygiene conditions for raw-milk processing, and the delivery of pasteurized milk to the sale unit.

Table 2 lists the prioritized action points from the workshops. The table shows top-five action points per chain participant and hazard. Relative importance is derived from respondents' utilities for action points and related control measures, and expressed as percentages (Valeeva et al. 2005). Numbers illustrate that none of

Table 2. *Important precautionary action points along the supply chain of fluid pasteurized milk, their relative importance, and third-party verifiability of related control measures*

Precautionary action points	Relative importance (%)[1]	Third-party verifiability[2]
Feed company – chemical hazards (k=21; n=4)		
Procedures and instructions for compound feed production	6.84	Fully
Quality assurance system of feed ingredient manufacturers	6.74	Fully
Adequate cleaning and disinfection of production equipment and premises	5.97	Fully
Finished compound-feed identification and traceability	5.81	Fully
Feed ingredients identification and traceability	5.71	Fully
(Total importance of top-five precautionary action points)	*31.07*	
Feed company – microbiological hazards (k=23; n=4)		
Feed ingredient identification and traceability	9.08	Fully
Quality assurance system of feed ingredient manufacturers	6.03	Fully
Adequate cleaning and disinfection of compound-feed transport vehicles	5.37	Fully
Adequate conditions of feed ingredients storage and intake	5.34	Fully
Quality assurance system of feed ingredient carriers	5.09	Fully
(Total importance of top-five precautionary action points)	*30.91*	
Dairy farm - chemical hazards (k=13; n=12)[3]		
Identification of treated cows in milking parlour	11.22	Fully
Quality assurance system of compound-feed manufacturers	10.50	Partly
Action in case of doubt about the withdrawal period	9.10	Fully/Partly
Origin of forage	8.32	Partly
Best farm practices	7.79	Partly
(Total importance of top-five precautionary action points)	*46.93*	
Dairy farm – microbiological hazards (k=30; n=13)		
Manure supply source	4.40	Partly
Action in salmonellosis and *M. paratuberculosis* cases	4.35	Fully
Acquisition of cattle	4.27	Fully
Udder cleaning before milking	4.16	Not/Partly
Calves feeding before weaning	4.11	Not/Partly
(Total importance of top-five precautionary action points)	*21.29*	
Dairy industry – chemical hazards (k=7; n=5)		
Sourcing raw milk	19.27	Fully
Delivered raw-milk identification and traceability	18.63	Fully
Procedures and instructions for raw-milk processing	14.30	Fully
Finished-product identification and traceability	12.89	Fully
Water used for production purposes	12.64	Fully
(Total importance of top-five precautionary action points)	*77.73*	
Dairy industry – micro-biological hazards (k=23; n=5)		
Finished product identification and traceability	5.97	Fully
Location of sealing equipment	5.96	Fully
Maintenance of the equipment and leakage prevention	5.62	Fully
Adequate cleaning and disinfection of raw-milk collection vehicles	5.21	Fully
Sourcing raw milk	5.09	Fully
(Total importance of top-five precautionary action points)	*27.85*	

[1]For each chain participant and group of hazards, there were *k* action points and *n* respondents. Per *k* action points, importance figures add up to 100%.

[2]Assuming no fraud.

[3]One respondent was removed from the analysis due to a mistake made in the validation profiles.

the top-fives embraces more than 50% of total importance, which equals 100% for each chain participant and group of hazards. (An exception to this is the top-five for chemical hazards at dairy-industry level, which embrace 77.7%, but this is due to the low number of action points, i.e., 7.). These numbers illustrate that many of the 82 precautionary action points are perceived as important in preventing against chemical and microbiological hazards.

Our findings, i.e., a long list of relevant precautionary action points for only two perils, imply that specifying rules of behaviour and classifying the insured based on a number of separate precautionary action points may not be feasible. Instead, insurers might consider 'packages' of measures such as already existing quality assurance schemes demanding defined protocols for a large number of a company's processes.

A RISK ASSESSMENT FRAMEWORK FOR CALCULATING PREMIUM RATES

For a number of dairy supply-chain stages and processes, Table 3 shows the estimated size of batches and related processing times, as well as the calculated lost values of destroyed products and the costs of handling and notifications.

Batches are not specified for all production processes since for our case studies we assume a contamination, or cross-contamination, only to occur in a few stages and processes: through feed ingredients (leading to multiple 4-ton processing batches to be contaminated), the storage of feed at farm level (in silos of on average 14 tons), the storage of raw milk at farm level (in tanks of on average 5 tons), the collection of milk (in trucks of on average 20 tons) and the storage of milk at industry level (in tanks of on average 150 tons). *Average* values of batches and related processing times are used for tracking products forward along the chain. For instance, if a contamination of raw milk at the farm level is notified 2 hours after delivery, we assume that the milk is still in the collection vehicle, not yet in the storage tanks of the dairy industry. Also, in case of tracking the number of farms having received contaminated compound feed this number is based on the amount of feed produced and the average storage capacity of feed at the farm level. *Maximum* batch sizes and processing times listed in Table 3 are used for tracing products back into the chain to identify all suspected product. For instance, if a dairy farm encounters problems with compound feed, the feed supplier's production of the past 336 hours (14 days) becomes suspected. In relation to the product values specified, it is assumed that recalled products must be destroyed. Costs per kg of milk increase from Euro 0.31 per kilo at the farm level to Euro 0.69 at retail level. Notification costs only include media-announcement costs once products reach consumer level.

Table 3. *Framework for risk assessment*

Chain stages, processes and related products	Batch (1000 kg)[1]	Related time (hours)[1]	Recall expenses		
			Product[2] (Euro/kg)	Handling[3] (Euro/kg)	Notification[4]
Feed company					
Processing (feed)	4	8 (4; 24)	0.15	0.25	-
Transport (feed)	*	2 (1; 4)	0.20	0.25	-
Dairy farm					
Storage (feed)	14 (8; 24)	168 (2; 336)	0.20	0.30	-
Storage (raw milk)	5 (3; 7)	36 (1; 72)	0.31	0.15	-
Dairy industry					
Collection (raw milk)	20 (10;33)	3 (1; 6)	0.32	0.15	-
Storage (raw milk)	150 (100; 400)	10 (1; 24)	0.34	0.15	-
Processing (processed milk)	*	1.5 (1; 2)	0.46	0.15	-
Packaging (processed milk)	*	1.5 (1; 2)	0.54	0.18	-
Transport (processed milk)	*	3 (1; 5)	0.59	0.18	-
Retail					
Storage (processed milk)	*	8 (1; 12)	0.61	0.20	-
Pickup (processed milk)	*	4 (1; 12)	0.66	0.20	75,000[5]
Retail (processed milk)	*	12 (2; 72)	0.69	0.20	75,000[5]

[1]Average value, minimum and maximum between brackets.
[2]Products are destroyed, i.e., salvage value of zero.
[3]Only transportation costs (local) and costs of destruction (full costs, no additional revenues).
[4]Only media costs; as soon as products reach the consumer level, recalls must be announced in the media.
[5]Worst case, i.e., three front-page announcements in major newspapers.
*These (much smaller) batches are not specified in this paper.

Table 4 presents a short description of the case studies. The second column of the table shows related products (feed, raw milk, processed milk) and, if clear, an indication of whether these products are at the customer level or further along the chain. The third column includes estimated recall expenses. Case studies refer to (1) 400 tons of contaminated compound feed; (2) 5 tons of contaminated raw milk; (3) 150 tons of contaminated processed milk; and (4) 1 can of contaminated milk at retail level for which the source of contamination cannot be readily detected. Cases (1) to (3) refer to tracking products forward along the chain. They are varied with respect to the promptness of the product recall. Case (4) refers to a situation of both tracking and tracing.

Table 4. *Case studies risk assessment[1]*

Description	Products[2]	Recall expenses (1,000 Euro)[3]
1. *400 tons of contaminated feed[4]*		
1a. Recall is announced 1 day after delivery. Tracking the feed leads to 30 dairy farms. These farms still have 98% of the feed in their silos and none of the farms has delivered milk yet. All stored milk is destroyed.	Feed[a] Raw Milk[a]	200 35 *235 (total)*
1b. Recall is announced 3 days after delivery. 80% of the feed is still in farm silos. 15 farms have not yet delivered any milk. All stored milk is destroyed. The other 15 farms have already delivered milk to various dairy companies. This involves 4 milk collection vehicles and 2 storage tanks, 1 of which has been processed until packaging and 1 until retail. All this milk is recalled and destroyed as well.	Feed[a] Raw milk[a] Processed milk[b]	160 32 315 *507 (total)*
2. *5 tons of contaminated raw milk*		
2a. Contamination is detected just after delivery. The milk is still in the collection vehicle.	Raw milk	*9.5 (total)*
2b. Contamination is detected after 3 days. The milk was delivered 2 days ago. All the milk went into 1 collection vehicle and 1 storage tank, and is now at retail level. Dairy industry recalls the whole batch. Retail agrees with recalling a single batch and does not remove other batches or products from the shelves.	Processed milk[b]	*210 (total)*
3. *150 tons of contaminated processed milk*		
3a. Contamination is detected just after delivery. The milk is not yet stored in retail.	Processed milk	*135 (total)*
3b. Contamination is detected after 3 days. 50% of the milk is still at retail level; the other 50% has already been sold. Consumers are notified to return purchased cans.	Processed milk[a/b]	*210 (total)*
4. *A retailer finds a can of contaminated milk, produced 2 days ago. The source of contamination cannot be readily detected.*		
2 storage tanks of the related dairy company become suspected. All this milk is at retail level and needs to be recalled. In addition, 60 dairy farms become suspected including their delivered and stored milk of the last 3 days. Stored milk (20%) is destroyed. Delivered milk (80%) is tracked to 6 storage tanks, 50% of which have been processed until packaging and 50% until retail. All this milk is recalled as well, which requires a second media announcement. Also, 2 feed companies become suspected, including their feed production of the past 14 days, implying further (announcements of) recalls of feed and milk. 50% of the related milk has already been consumed; 40% is still at farm level; 10% causes raw-milk destruction and processed milk recalls (compare 1b).	Processed milk[a] Raw milk Processed milk[a] Feed Raw milk[a] Processed milk[b]	340 26 800 800 32 315 *2,313 (total)*

[1]Based on (average) batch, time and cost parameters of Table 3.
[2]Superscript characters refer to the scope of losses (if clear): 'a' relates to products at *customer* level, 'b' to products *further along the chain.*
[3]Value of destroyed products and costs of handling and notification.
[4]Equals compound feed production of average cattle feed factory *per day* (100,000 per year).

Cases studies illustrate that even non-compliance of single batches, or even a single product, can lead to considerable losses. Also, it is illustrated that late recalls lead to higher losses than early recalls (1b-1a, 2b-2a, 3b-3a). In addition, the fourth case shows that recall losses can easily cumulate. Cases also demonstrate the importance of strictly defining perils and losses of a product recall, as discussed earlier in this paper. For instance, in case 1b, largest losses occur 'further along the chain' – not at customer level. Also, even in situations 2a and 2b neither of the losses is at customer level. In the fourth case, in which the source of contamination is not readily detected, defining the scope of losses is even more difficult. Note that with case 2a, a product recall insurance strictly covering recalls at customer level could yet give the wrong incentives, i.e., to postpone the product recall until the milk is stored in dairy industry storage tanks. As such, the framework and case studies presented provide a useful basis for a structured analysis of recall losses, as a basic element for insurers to estimate adequate premium levels.

THIRD-PARTY VERIFIABILITY OF DUE DILIGENCE

Insurance payments generally cover 'accidental and unintentional losses' (Rejda 1992). In order to verify that the insured followed 'proper rules of behaviour', it is useful for insurers to being able to check the insured's due diligence. Terms are clarified by the following example (derived from Blanchfield (1992) and Schothorst and Jongeneel (1992)): In order to avoid the risk of crossing red traffic lights, a *precautionary action point* relates to the brakes. A relevant *control measure* would be 'brakes in working order'. Then, *due diligence* consists of regular checking that the brakes are indeed in satisfactory condition. This due diligence would be *verifiable* by a third party if (1) the checking is validated to give good insight into the working of the brakes, and (2) the results were registered.

Due diligence thus relates to the proper application of an adequate control measure, and third-party verifiability refers to being able to demonstrate objectively that this proper application is ensured.

The precautionary actions points (Table 2) and their related control measures were validated in literature and through expert evaluations. We now focus on the ability to provide objective evidence of ensuring their proper application. Assuming that there is no fraud, Table 2 (last column) shows that most control measures can be fully verified by a third party, especially those at the feed and dairy industry level. However, at the farm level, not all control measures are fully verifiable. For instance, with respect to 'best farm practices', it is fully verifiable whether there is adequate cleaning equipment, sufficient disinfectants and an adequate level of training of the farmer and his employees, but it is not verifiable whether disinfectants are properly used and people always work according to hygiene rules. For these reasons, 'best farm practices' were assessed to be only partly verifiable. Also, the origins of forage and compound feed were assessed as partly verifiable; although auditors can check accounts of goods purchased, it is not verifiable whether they represent all of the present goods. The same applies to the manure supplied to a farm.

A fully verifiable measure at dairy farms includes, i.e., the 'identification of treated cows in the milking parlour'. To verify whether farmers identify treated cows in the milking parlour in order to discard the milk of these cows, auditors can check the prescriptions of veterinarians. The same applies for the 'action in salmonellosis and *M. paratuberculosis* cases'. Also, 'acquisition of cattle' is a fully verifiable action point since there is a well-functioning identification and registration system. The 'action in case of doubt about the withdrawal period' is only fully verifiable if farmers consistently carry out tests on their milk; not if they just extend the withdrawal period with a few days.

A farm-level top-five precautionary action point which was assessed as 'not/partly' verifiable is the 'udder cleaning before milking'. Continuous monitoring of the control measures related to this action point, e.g., 'wet cleaning' or 'cleaning with dry towel – one towel per cow' is not reasonable (in case of traditional milking systems). Also, even unannounced hygiene audits would not lead to full verifiability as 'udder cleaning before milking' concerns a typical handling which can easily be changed into desired behaviour during an audit. The same issues apply to the 'calves feeding before weaning': do calves always get milk from a non-suspected cow or are they fed with milk from some arbitrary cow?

At feed and dairy industry level many of the issues are more formalized and therefore better verifiable by third parties. For instance, identification and traceability can be verified through records and coding, adequate cleaning is regularly checked upon and sourcing of ingredients can be checked not only through accounts of goods purchased but through entire input/output balances as well.

In conclusion, for feed and dairy industries all of the top-five precautionary action points and related control measures are fully verifiable by a third party. However, at the farm level processes are less intensively monitored, and, therefore, difficulties occur with respect to the third-party verifiability of farmers' due diligence.

CONCLUSIONS

Some insurers provide cover for product recall, others do not (anymore). In this paper we started out by stating that changes in the risk environment of food supply chains lead to insurers facing designing and premium rating problems with food-related products such as recall insurance. Through our four objectives we disentangled these problems for recall insurance schemes in dairy supply chains into four issues: (1) a proper definition of perils and losses of product recall; (2) an identification of important precautionary action points for underwriting, class rating and specifying 'proper rules of behaviour'; (3) a framework for risk assessment and rating; and (4) an assessment of third-party verifiability of due diligence for identifying the insured's eligibility for indemnification. From the analyses we conclude that:

- Perils and losses of product recall need to be strictly defined, preferably on a case-by-case basis in order to prevent pitfalls of ambiguity and to keep incentives for risk prevention straight.
- Underwriting, class rating and rules of behaviour should, if possible, be linked to already existing quality assurance programmes as there are probably too many

relevant precautionary action points and related control measures for insurers to assess.

- Recall losses easily accumulate as we already identify losses between Euro 210,000 and Euro 2,300,000 for contamination problems in single batches and products, and considering only a few recall expenses.
- Third-party verifiability of due diligence is somewhat problematic at the farm level, mainly because of less intensive monitoring systems. Since this issue is not easy to solve, i.e., because of the relatively small scale of most farms, insurers will have to look for other tools providing 'due-diligence incentives' to farmers, such as deductibles and co-payments.

Although our analyses focused on dairy supply chains, we believe these issues to be applicable to other food supply chains as well. Overall, we conclude that, if risks are properly defined and insurance schemes incorporate adequate incentives for due diligence and rapid disclosure, product recall is an insurable type of risk, even if the number of recalls will further increase in the coming years.

ACKNOWLEDGEMENTS

The authors would like to thank the Dutch insurance company Achmea, the Dutch Dairy Organization (NZO) and the Mesdagfonds for funding the study on precautionary-action points. Also, enthusiastic cooperation of farm and industry experts throughout the whole project is greatly appreciated.

REFERENCES

Arrow, K.J., 1996. The theory of risk-bearing: small and great risks. *Journal of Risk and Uncertainty,* 12 (2/3), 103-111.

Blanchfield, J.R., 1992. Due diligence: defence or system? *Food Control,* 3 (2), 80-83.

CBS and LEI, 2005. *Agricultural and horticultural data 2005.* Cental Statistical Bureau, Voorburg.

Collins, M.T., 1997. *Mycobacterium paratuberculosis:* a potential food-borne pathogen? *Journal of Dairy Science,* 80 (12), 3445-3448.

Cullor, J.S., 1995a. Common pathogens that cause foodborne disease: can they be controlled on the dairy? *Veterinary Medicine,* 90, 185-194.

Cullor, J.S., 1995b. Implementing the HACCP program on your clients' dairies. *Veterinary Medicine,* 90 (3), 292-295.

Cullor, J.S., 1997. HACCP (Hazard Analysis Critical Control Points): is it coming to the dairy? *Journal of Dairy Science,* 80 (12), 3449-3452.

Gould, B.W., Smukowski, M. and Bishop, J.R., 2000. HACCP and the dairy industry: an overview of international and U.S. experiences. *In:* Unnevehr, L.J. ed. *The economics of HACCP: costs and benefits*. American Association of Cereal Chemists, St. Paul, 365-384.

Henson, S. and Hooker, N.H., 2001. Private sector management of food safety: public regulation and the role of private controls. *International Food and Agribusiness Management Review,* 4 (1), 7-17.

Mathews, K.H., Buzby, J.C., Tollefson, L.R., et al. 2001. Livestock drugs: more questions than answers? *Agricultural Outlook* (September), 18-21 [http://www.ers.usda.gov/publications/AgOutlook/sep2001/ao284g.pdf]

Meuwissen, M.P.M., Velthuis, A.G.J., Hogeveen, H., et al. 2003. Traceability and certification in meat supply chains. *Journal of Agribusiness,* 21 (2), 167-181.

Novoselova, T.A., Meuwissen, M.P.M., Van der Lans, I.A., et al. 2002. *Consumers' perception of milk safety: 13th congress of the International Farm Management Association (IFMA), 7-12 July 2002, Papendal Conference Center, Arnhem, The Netherlands*. IFMA, Cambridge. [http://www.ifmaonline.org/pdf/congress/Novoselova.pdf].

Rejda, G.E., 1992. *Principles of risk management and insurance*. Harper Collins, New York.

Segerson, K., 1999. Mandatory versus voluntary approaches to food safety. *Agribusiness New York,* 15 (1), 53-70.

Sischo, W.M., Kiernan, N.E., Burns, C.M., et al. 1997. Implementing a quality assurance program using a risk assessment tool on dairy operations. *Journal of Dairy Science,* 80 (4), 777-787.

Skees, J.R., Botts, A. and Zeuli, K.A., 2001. The potential for recall insurance to improve food safety. *International Food and Agribusiness Management Review,* 4 (1), 99-111.

Teratanavat, R. and Hooker, N.H., 2004. Understanding the characteristics of US meat and poultry recalls: 1994-2002. *Food Control,* 15 (5), 359-367.

Valeeva, N.I., Meuwissen, M.P.M., Lansink, A.G.J.M.O., et al., 2005. Improving food safety within the dairy chain: an application of conjoint analysis. *Journal of Dairy Science,* 88 (4), 1601-1612.

Van Schothorst, M. and Jongeneel, S., 1992. HACCP, product liability and due diligence. *Food Control,* 3 (3), 122-124.

Veling, J., Wilpshaar, H., Frankena, K., et al., 2002. Risk factors for clinical *Salmonella enterica* subsp. *enterica* serovar Typhimurium infection on Dutch dairy farms. *Preventive Veterinary Medicine,* 54 (2), 157-168.

CHAPTER 12

STRATEGIC ALLIANCES AND NETWORKS IN SUPPLY CHAINS

Knowledge management, learning and performance measurement

THOMAS L. SPORLEDER

Agribusiness and Farm Income Enhancement Endowed Chair, Department of Agricultural, Environmental, and Development Economics, The Ohio State University, 2120 Fyffe Road, Columbus, Ohio 43210. E-mail: sporleder.1@osu.edu

Abstract. This manuscript defines and analyses the concept of a strategic alliance as one specialized collaborative agreement among vertically-allied firms in the supply chain. Vertical relationships and alliances coagulate among upstream and downstream firms in an effort to form networks that are synergistic and add value beyond what an individual firm may be able to achieve. One driver to form a strategic alliance is intellectual property that serves as a base for *maximizing value added* within a supply chain. Multiple diverse organizations that collaborate within a supply chain compose a network.

Knowledge management is introduced in the analysis of strategic alliances. Knowledge management logic helps in understanding the information-sharing aspects of a strategic alliance. Ambiguity plays a role in the extent to which information is shared. Thus, knowledge management provides novel insight into the foundations of a strategic alliance. The potential of a strategic alliance creating a real option for managers is examined along with the characteristics of networks that are organized around constant learning.

Strategic-alliance performance evaluation also is addressed. Sometimes it is not appropriate to evaluate the strategic alliance based on conventional means such as profit and return on investment. Strategic alliances may involve objectives such as entering new markets, learning and obtaining new skills, and/or sharing risks and resources. When a profit centre is not part of the object of cooperation the alliance presents challenges to managers in terms of evaluation. Performance evaluation of alliances is suggested based on a certain-to-fuzzy continuum of inputs and outputs.
Keywords: supply-chain performance; resource-based theory; agribusiness; food

KNOWLEDGE MANAGEMENT AND PERFORMANCE MEASUREMENT OF STRATEGIC ALLIANCES IN FOOD SUPPLY CHAINS

The globalization of the food system has been rapid and resulted from numerous factors. Among those factors are lessening national boundaries through freer trade, and rapid technological advance in areas such as biotechnology, communication and information technologies, and transportation and packaging technologies. The past decade has witnessed genetically engineered commodities, global positioning

C.J.M. Ondersteijn et al. (eds.), Quantifying the agri-food supply chain, 161-171.

systems for production agriculture, cheaper and better computers, and aseptic packaging, which allows cost-effective shipment of relatively low-value commodities over long distances (Sonka et al. 2000). At the same time, domestic trade policy provided enhanced free-market incentives and encouraged firms to reach beyond their traditional geographic perspectives (Sporleder and Martin 1998).

Strategic partnering among firms is one response to this more challenging and complex environment. Partnering among firms may take numerous forms, ranging from informal alliances to more formal joint ventures (Harrigan 1988). The purpose of this manuscript is to examine drivers underlying managerial decision-making regarding firms entering into strategic alliances, where joint ventures are regarded as part of the broader definition of strategic alliances. Strategic partnering is one of a broader class of governance structures that may be useful in achieving enhanced vertical coordination in the supply chain.

The emerging area of knowledge management is introduced in the analysis of strategic alliances. Knowledge management helps in understanding a firm's willingness to enter into strategic partnering with another firm where the object of co-operation cannot be evaluated using conventional means. The structure of knowledge management is useful in providing novel pathways in which to explore interfirm information sharing. Knowledge management logic is especially useful by providing additional characteristics of a strategic alliance, such as the potential for learning and creating managerial flexibility. Such characteristics provide novel insight into incentives for entering into strategic alliances among vertically-allied economic agents within a supply chain.

Strategic alliances are viewed as a special case of strategic partnering. The analysis specifically focuses on the issue of performance evaluation of strategic alliances, especially when there is no separate profit centre created as part of the alliance. If no profit centre is a part of the object of cooperation, performance evaluation becomes more arduous and complex. In this situation, the partners to the alliance typically cannot use conventional performance measures, such as profit or return on investment, to judge the performance of the alliance or to evaluate the wisdom of their partners' decision to enter into the alliance.

ALTERNATIVE EXCHANGE MECHANISMS

Alternative exchange mechanisms may be categorized based on the relative extent of vertical control available from the mechanism (Sporleder 1992). Broad alternatives are spot markets, contracts, strategic alliances (including joint ventures) and ownership integration (Peterson and Wysocki 1998). The extremes of the continuum are the spot-market alternative, which offers virtually no vertical control, while vertical ownership integration provides the firm with relatively strong vertical controls through ownership of another stage or industry within the vertical chain. Contracting and strategic alliances offer increasing vertical control relative to spot markets, but the negative is increased idiosyncratic investment by the firm.

There are several strategic partnering options available to firms participating in the global food system. Strategic partnering involves a broad class of activities.

Contracts, strategic alliances and ownership integration are the three most basic forms of strategic partnering. Interestingly, Peterson and Wysocki (1998) propose a choice model that managers might employ to decide about one coordination strategy over another.

CHARACTERISTICS OF STRATEGIC PARTNERING ALLIANCES

Strategic alliances are a form of strategic partnering, but partnering also includes contracting, ownership integration, and/or entering into mergers and consolidations. Performance evaluation of strategic alliances is of particular concern in this manuscript. However, before turning to this issue, the specific types of strategic alliances are categorized and the characteristics of the various strategic partnering alternatives are identified.

Types and characteristics of strategic alliances

Strategic alliances are defined as any agreement between or among firms to cooperate in an effort to accomplish some strategic purpose. The categorization of strategic alliances is based on Barney (2002) and captures the essence of contemporary thought regarding strategic alliances. Categorization includes three types of strategic alliances: non-equity alliances, equity alliances and joint ventures. In non-equity alliances, each firm to the agreement is a stakeholder, but not necessarily a shareholder in the object of the cooperation. By contrast, equity alliances and joint ventures typically are a more formal configuration for a strategic alliance where the partners become both stakeholders and shareholders, in the sense that the partners contribute equity capital to the joint venture. Also, typically the resultant object of cooperation (often a newly-defined business) is operated as a profit centre.

Non-equity alliances represent cooperation between firms, managed less formally than the other forms of strategic alliances. Sporleder (1994) has articulated distinguishing factors unique to non-equity strategic alliances, including fuzzy prerogatives and fuzzy obligations relative to joint ventures, relatively weak and malleable vertical control, and partners which are stakeholders in the object of the alliance but not necessarily shareholders. Rarely is a new independent firm created. Trust is a cornerstone of these less formal and often fuzzy arrangements[1].

By contrast, equity strategic alliances and joint ventures refer to business relationships where agreements are supplemented by equity investments by one partner in the other, an action that is often reciprocated. These types are more formal, involve capital investment, and consequently the partners to the arrangement become shareholders as well as stakeholders[2]. Joint ventures are distinguished from equity strategic alliances as cases where firms agree to cooperate with each other to achieve a specific, relatively well-defined, goal. The participating companies usually form a new and separate legal entity in which they invest. Typically, profits from the joint venture provide compensation for the partners (Kogut 1988).

Major stimuli for food processors entering into a strategic alliance with their suppliers include (in the order of importance) cost control, developing product prototypes, improving product quality, and improving package design (Food Processing Magazine). Over one-fourth of the alliances were formed for reasons of cost control while another 45% were formed for R&D purposes of improving existing product formulations or developing new products. Food processors are consistent with general manufacturing firms in joining strategic alliances primarily for the purpose of improving operational efficiency or learning and technology transfer.

KNOWLEDGE MANAGEMENT AND STRATEGIC ALLIANCES

Knowledge management has emerged recently as an integrated approach to identifying, creating, managing, sharing and exploiting the information and knowledge assets of an organization (Sporleder and Moss 2002). The importance of skill acquisition, learning and the accumulation of capability over time is the core of knowledge management within an organization (Nonaka 1994; Teece 2000). Organizational knowledge management may be viewed as a process of knowledge creation and the organizational performance outcomes that result from that knowledge (Soo et al. 2001). Information sources include networks for acquiring information from internal and external sources. The notion is that networking improves the flow of information.

Learning

Learning capacity differs among firms or agents in the supply chain. The absorptive capacity (learning capability) of an individual or organization is the ability to recognize, assimilate and incorporate information, either internal or external to the organization (Cohen and Levinthal 1990). Absorptive capacity partially determines the use of knowledge and the quality and scope of decision-making based on it. One tenet of the model is that that as absorptive capacity of an organization or an individual improves, the more new knowledge is created (Powell et al. 1996). The knowledge management logic is based on the notion that knowledge creation is positively correlated with both innovation (Nonaka 1994) and financial performance (Nelson and Winter 1982). Innovation and improved performance are the end points from new organizational knowledge.

The application of knowledge management logic to strategic alliances seems appropriate. One driver behind the formation of strategic alliances is often regarded as information sharing or exchange (Sporleder 1994). The aspect of knowledge transfer in strategic alliances is focused on causal ambiguity that is common in resource-based theory of the firm. Ambiguity conceptually provides barriers to imitation, which makes it difficult for rivals to know which competencies form the basis for competitive advantage (Simonin 1999). Ambiguity is empirically verified by Simonin (1999) to play a major role in the knowledge transfer process among alliance members. Thus, ambiguity is a contingency that appears to influence the outcomes of knowledge transfer in a strategic alliance. Ambiguity joins the list of other factors thought to influence knowledge transfer such as complementarities of existing firm assets among alliance partners and the governance mechanism employed by the alliance. Complementarities of assets are thought to enhance the firm's capacity to understand new information from the partners of the alliance.

Opportunism and trust are thought to be important in the outcome of a strategic alliance. The extent of trust is rooted in the cultural-value similarities among alliance members and may be related to the social capital of the organizations of the alliance. This social-capital direct tie back to the knowledge management literature could serve as the base for numerous interesting and novel hypotheses and interactive influences regarding information sharing, trust and social capital in alliances.

Real options

Finally, the notions of relational embeddedness and structural embeddedness flowing from knowledge management logic may be important to understanding why strategic alliances form among particular firms and not others. Network embeddedness, encompassing both structural and relational embeddedness, may influence the outcome of a firm's participation in an alliance and could affect the design and implementation of strategy relating to quality signalling in supply chains (Sporleder and Goldsmith 2001). The type of social capital that generates a competitive advantage over rivals may depend on the competitive environment. Firms engaged in knowledge exploitation, rather than exploration, may require specific knowledge that is best procured from dense network structures (Rowley et al. 2000). However, dense networks may cause firms to neglect or not fully appreciate new information and alternatives (Nahapiet and Ghoshal 1998).

PERFORMANCE EVALUATION OF STRATEGIC ALLIANCES

Numerous analysts have written about evaluation of joint ventures. Some analysts have noted an apparent long-term instability of joint ventures, both international and domestic. Empirical studies concerning the instability of joint ventures often use proxies for instability. Blodgett (1992) used renegotiation of the venture contract or any change in equity division as a proxy for instability. Inkpen and Beamish (1997) used a change in partner relationship or bargaining power to represent instability. Consensus among analysts is that strategic alliances, in general, are relatively unstable business arrangements even when there is a separate legal entity involved.

The performance of non-equity alliances is difficult to measure because there is no single 'indicator' of performance, such as profit/loss, that can be assessed. The role of management may be critical in these agreements. Non-equity alliances are transitional compared to other alternatives for strategic partnering. Evaluation of such alliances may evolve as a negotiated item between the partners.

There are several challenges related to evaluating joint ventures. Often, joint ventures are evaluated as if they were a division of the parent (Anderson 1990). This method of evaluation may cause dissonance relative to which parent performs the evaluation. Another challenge is that joint ventures may not receive an accurate evaluation if they are evaluated in the same manner as a wholly-owned division of the parent. A joint venture is a shared entity and unless the method of evaluation is specified this might cause some conflict (Pearce 1997). The goals of the parents and the joint venture may be divergent, so evaluating the joint venture as a division may not be optimal. Although it might be easier to determine profitability and other standard performance measures, joint ventures may be deployed in risky, uncertain situations with high levels of instability. Thus, the sole criteria of profitability might not provide an accurate account of how the joint venture is performing.

In reality however, without a profit centre the financial aspect of performance evaluation may not be possible. The focus, therefore, is on the relationship among the alliance partners as well as on the resources devoted to the alliance by each partner. A firm that partners in an alliance may evaluate its own performance after engaging in an alliance and may be able to ascertain the impact of the relationship on its own profitability. Evaluating the alliance, the relationship or agreement between the companies, however, may remain a point of obscurity.

Alliance evaluation criteria based on a certain-to-fuzzy continuum

The role of management is critical when evaluating strategic alliances. They need to be aware of what types of resources, tangible and intangible, are dedicated to the strategic alliance. According to traditional methods, a manager may be required to determine performance based on the amount of stockholder equity to debt that is held by the company, the level of profitability of a company, the productivity (i.e., output per hour), or even participation in the global market. However, conventional output measures may be sufficient for, or even relevant to, performance evaluation, especially in the case of a non-equity strategic alliance.

The concept of using weights in evaluating joint ventures refers to how heavily inputs and outputs should be considered in the process. For example, should learning be given more importance than marketing performance? Following Ouchi (Ouchi 1979) the first dimension examines how certain managers are regarding how inputs become outputs – the transformation process. The second dimension encompasses the extent to which a firm is able to assess measure and judge results (outputs). A combination of these two dimensions results in the analytic framework of Figures 1 and 2. The generic space defined in Figure 1 simply provides the analytic framework for determining the relative performance evaluation outcomes for non-equity strategic alliances, equity strategic alliance, and joint venture.

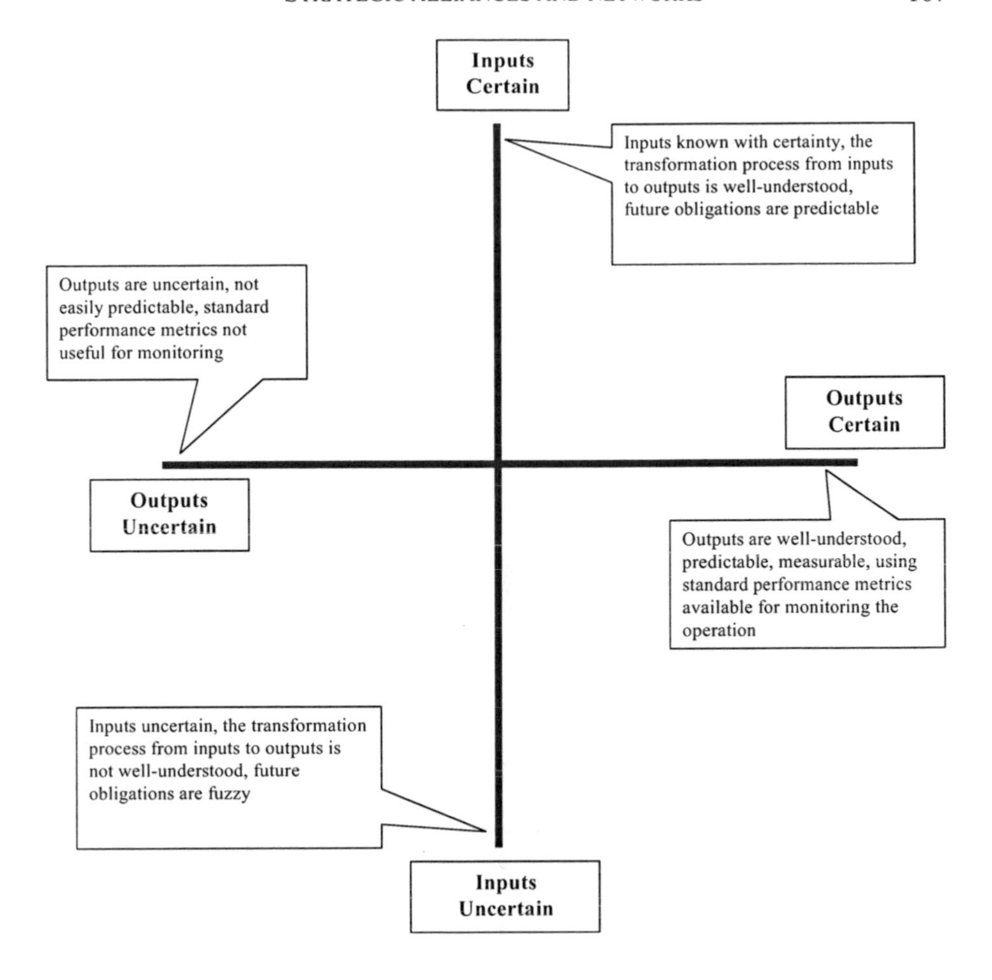

Figure 1. *Relative space for strategic-alliance performance evaluation, based on a certain-to-uncertain continuum inputs and outputs*

On the right side of the output continuum of Figure 1 managers have a poor grasp of the transformation process; therefore, outputs cannot be accurately assessed. The northwest quadrant of Figure 1 represents strategic-alliance cases where the transformation process is well-understood but outputs cannot be accurately assessed. In this case, input measures are heavily weighted and output measures weighted lightly. In the southeast quadrant of Figure 1 managers have a poor understanding of the input-output process but are able to assess outputs with some certainty. In this southwest quadrant case, output measures are heavily weighted and input measures weighted lightly.

The northeast quadrant represents the ideal case where the partners in the alliance realize what inputs to contribute and are able to evaluate outputs with accuracy. In this quadrant the use of both inputs and outputs in the evaluation process is valid. Both variables should be used with more weight being placed on outputs because these measures of performance can be obtained and evaluated.

Evaluating non-equity alliances

The contributions of Ouchi (1979) and Anderson (1990) provide the foundations for a model adapted to exclude financial (results-oriented) methods of evaluations but to emphasize the input variables that indicate the state of an alliance. The northwest and southwest quadrants of Figure 2 were used appropriately as a guideline for evaluating these alliances.

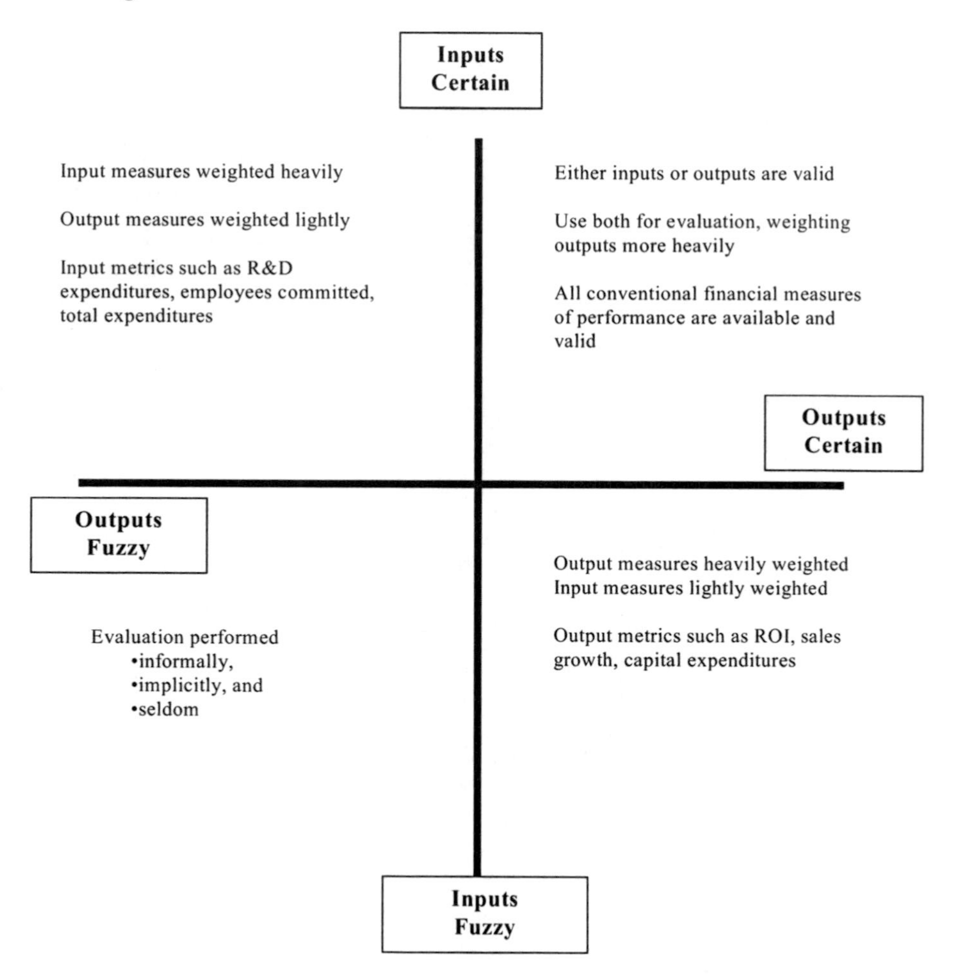

Figure 2. *Relative space for strategic-alliance performance evaluation, based on a certain-to-fuzzy continuum inputs and outputs*

The southwest quadrant represents alliances operating in information poor situations. This is usually the initial stage of a relationship where firms are operating with fuzzy prerogatives. The appropriate actions and inputs are often unclear to the partners of the alliance. It might also be the case that the firms are unsure of what the outputs of the relationship will be or should be, hence the perplexity of the transformation process.

The northwest quadrant refers to cases where managers have a better grasp on what they should be doing and what actions should be taken to meet the objectives of the alliance. However, there is still no clear means for assessing outputs. This is where proper definition of the goals and objectives of the alliance becomes important.

Non-equity alliances typically are placed in either the northwest or northeast quadrant. Since these alliances furnish no standard results-oriented measures of evaluations, the focus then turns to inputs variables, classified by Anderson (1990) as states of being. States of being refers to how the alliance is doing. Is there harmony among the alliance partners? Is there high morale among the employees of the company? Are there sufficient levels of communication between the alliance partners to facilitate a successful relationship? There are no standard measures of harmony and the presence of conflict might indicate a lack of harmony. The evaluation of strategic alliances rests heavily on the managers' shoulders. It relies on managers' abilities to understand the inputs (i.e., human resources) that are necessary for the particular alliance, their ability to communicate with employees, and to motivate them to act accordingly. The input measures should be utilized with the goal or objective of the alliance in mind.

Evaluating non-equity strategic alliances is a subjective process because managers must decide what is working and what is not. Although the manager may be unable to evaluate the output, if there is some certainty regarding the necessary inputs the alliance can be evaluated in this manner.

It is important to note that resources dedicated to an alliance may be evaluated rather than more conventional output measures of performance. In the case of non-equity alliances, only inputs are likely to be evaluated. The role of managers is critical to performance evaluation of these alliances, because without measurable outputs managers can still provide some evaluation.

CONCLUSIONS AND IMPLICATIONS

Strategic alliances are agreements between or among firms to cooperate in an effort to accomplish some strategic purpose. Each firm to the agreement is a stakeholder, but not necessarily a shareholder, in the object of the cooperation. By contrast, joint ventures typically are more formal configurations for a strategic alliance where the object of cooperation is operated as a profit centre. Thus, performance evaluation of the partnership resulting from the alliance is through conventional means such as profit and return on investment. However, other types of strategic alliances may involve objectives such as entering new markets, obtaining new skills, and/or sharing risks and resources. If no profit centre is a part of the cooperation, performance evaluation becomes more arduous and complex. Methods of alliance evaluation are suggested.

Knowledge management is introduced in the analysis of strategic alliances. Knowledge management logic helps in understanding the information-sharing aspects of a strategic alliance. Ambiguity plays a role in the extent to which information is shared. Thus, knowledge management provides novel insight into the foundations of a strategic alliance.

Non-equity strategic alliances help in understanding a firm's willingness to enter into strategic partnering with another firm where the object of cooperation cannot be evaluated using conventional means. Non-equity strategic alliances, in general, are inherently different from either equity strategic alliances or joint ventures. Distinguishing factors, unique to strategic alliances, include fuzzy prerogatives and fuzzy obligations relative to joint ventures, relatively weak and malleable vertical control, and partners that are stakeholders in the object of the alliance but not necessarily shareholders. In the case of a non-equity alliance, only inputs are likely to be able to be evaluated. The role of managers is critical to performance evaluation of transitory alliances, although subjectivity and uncertainty are minimized.

NOTES

[1] Adams and Goldsmith (1999) provide a new analytic framework for fuzzy strategic alliances based on the codification of trust. Three levels of trust are explicitly recognized in their analysis. Their framework for fuzzy strategic alliances is enhanced by this perspective on trust.

[2] The equity structure among joint ventures is often 50:50 investments from its partners. There are, however, cases of minority/majority equity investments, such as 49:51, or some other agreed upon ratio.

REFERENCES

Adams, C.L. and Goldsmith, P.D., 1999. Conditions for successful strategic alliances in the food industry. *International Food and Agribusiness Management Review,* 2 (2), 221-248. [http://www.ifama.org/nonmember/OpenIFAMR/Articles/v2i2/221-248.pdf]

Anderson, E., 1990. Two firms, one frontier: on assessing joint venture performance. *Sloan Management Review,* 31 (2), 19-30.

Barney, J., 2002. *Gaining and sustaining competitive advantage.* Prentice Hall, Upper Saddle River.

Blodgett, L.L., 1992. Factors in the instability of international joint ventures: an event in history analysis. *Strategic Management Journal,* 13 (6), 475-481.

Cohen, W.M. and Levinthal, D.A., 1990. Absorptive capacity: a new perspective on learning and innovation. *Administrative Science Quarterly,* 35, 128-152.

Harrigan, K.R., 1988. Joint ventures and competitive strategy. *Strategic Management Journal,* 9, 141-158.

Inkpen, A.C. and Beamish, P.W., 1997. Knowledge, bargaining power, and the instability of international joint ventures. *Academy of Management Review,* 22 (1), 177-202.

Kogut, B., 1988. Joint ventures: theoretical and empirical perspectives. *Strategic Management Journal,* 9, 319-332.

Nahapiet, J. and Ghoshal, S., 1998. Social capital, intellectual capital, and the organizational advantage. *Academy of Management Review,* 23 (2), 242-266.

Nelson, R.R. and Winter, S.G., 1982. *An evolutionary theory of economic change.* Harvard University Press, Cambridge.

Nonaka, I., 1994. A dynamic theory of organizational knowledge creation. *Organization Science,* 5 (1), 14-37.

Ouchi, W.G., 1979. A conceptual framework for the design of organizational control mechanism. *Management Science,* 25 (9), 833-848.

Pearce, R. J., 1997. Toward understanding joint venture performance and survival: a bargaining and influence approach to transaction cost theory. *Academy of Management Review*, 22 (1), 203-225.

Peterson, H.C. and Wysocki, A.F., 1998. *Strategic choice along the vertical coordination continuum: paper prepared for the American Agricultural Economics Association Meetings, Salt Lake City, Utah, August 2-5, 1998*. Michigan State University, East Lansing. Staff Paper no. 98-16.

Powell, W.W., Koput, K.W. and Smith-Doerr, L., 1996. Interorganizational collaboration and the locus of innovation: networks of learning in biotechnology. *Administrative Science Quarterly*, 41 (1), 116-145.

Rowley, T., Behrens, D. and Krackhardt, D., 2000. Redundant governance structures: an analysis of structural and relational embeddedness in the steel and semiconductor industries. *Strategic Management Journal*, 21 (3), 369-386.

Simonin, B.L., 1999. Ambiguity and the process of knowledge transfer in strategic alliances. *Strategic Management Journal*, 20 (7), 595-623.

Sonka, S., Schroeder, R.C. and Cunningham, C., 2000. *Transportation, handling and logistical implications of bioengineered grains and oilseeds: a prospective analysis*. USDA Agricultural Marketing Service. [http://www.ams.usda.gov/tmd/LATS/LATSbiotech.PDF]

Soo, C., Devinney, T., Midgley, D., et al. 2001. *Knowledge management: philosophy, process, pitfalls, and performance*. INSEAD, Fontainebleau. Working Paper INSEAD.

Sporleder, T.L., 1992. Managerial economics of vertically coordinated agricultural firms. *American Journal of Agricultural Economics*, 74 (5), 1226-1231.

Sporleder, T.L., 1994. Assessing vertical strategic alliances by agribusiness. *Canadian Journal of Agricultural Economics*, 42, 533-540.

Sporleder, T.L. and Goldsmith, P.D., 2001. Alternative firm strategies for signaling quality in the food system. *Canadian Journal of Agricultural Economics*, 49 (4), 591-604.

Sporleder, T.L. and Martin, L., 1998. Economic perspectives on competitiveness under WTO, NAFTA and FTAA. *In:* Loyns, R.M.A., Knutson, R.D. and Meilke, K. eds. *Economic harmonization in the Canadian/U.S./Mexican grain-livestock subsector: proceedings of the fourth agricultural and food policy systems information workshop, April 22-25, 1998, in Lake Louise, Canada*. Friesen Printers, Winnipeg. [http://www.farmfoundation.org/blue/sporleder.pdf]

Sporleder, T.L. and Moss, L.E., 2002. Knowledge management in the global food system: network embeddedness and social capital. *American Journal of Agricultural Economics*, 84 (5), 1345-1352.

Teece, D.J., 2000. *Managing intellectual capital: organizational, strategic, and policy dimensions*. Oxford University Press, New York.

SUPPLY CHAIN ORGANIZATION AND CHAIN PERFORMANCE

CHAPTER 13

QUANTIFYING STRATEGIC CHOICE ALONG THE VERTICAL COORDINATION CONTINUUM

Implications for agri-food chain performance

ALLEN F. WYSOCKI[#], H. CHRISTOPHER PETERSON[##] AND STEPHEN B. HARSH[##]

[#] *Associate Professor University of Florida, 1161 McCarty Hall A, P.O. Box 110240, Gainesville, FL 32611-0240, USA. E-mail: wysocki@ufl.edu*
[##] *Professor Department of Agricultural Economics, Michigan State University*

Abstract. Given the increasing emergence of highly integrated agri-food supply chains, a key question arises as to how to measure the performance of these chains. This chapter postulates that agri-food supply-chain performance can be best understood with the help of three separate, but related phenomena: the individual firm's desire to participate in the supply chain; the governance structure of the whole chain; and the application of industrial organization and institutional economic theory. The Peterson, Wysocki and Harsh (PWH) model of vertical coordination strategy selection is provided. A brief overview of the various forms that agri-food chains may take on and a multi-disciplinary approach to understanding agri-food chain performance are offered, including channel master, chain web and chain organism. The chapter ends with research challenges still needing to be addressed including the limited access to information and measurement issues. We conclude that the PWH model, learning supply-chain governance structures, and application of additional economic theories model can be useful in understanding and measuring performance in agri-food chains.
Keywords: agri-food chain performance measures; chain web; chain organism; channel captain; integrated supply chain; learning supply chain; PWH model; unintegrated supply chain

INTRODUCTION

Given the increasing emergence of highly integrated agri-food supply chains, a key question arises as to how to measure the performance of these chains. Performance matters because policy makers care about individual and supply-chain performance. Firms within supply chains care about their firm-specific performance and the performance of other firms in a given supply chain, especially if their performance is impacted by others in the supply chain. Therefore, the question is, what would performance depend upon in agri-food supply chains?

C.J.M. Ondersteijn et al. (eds.), Quantifying the agri-food supply chain, 175-190.

This paper postulates that agri-food supply-chain performance can be best explained and understood with the help of three separate, but related phenomena: 1) the individual firm's desire to participate in the supply chain, 2) the governance structure of the whole chain, and 3) the application of industrial organization and institutional economic theory.

Coordination choices firms make along the vertical coordination continuum representing the various levels of supply-chain participation desired by firms are presented. The Peterson, Wysocki and Harsh (PWH) model of vertical coordination strategy selection is provided. A brief overview of the various governance structures that agri-food chains may adopt and a multi-disciplinary approach highlighting the relevance of industrial organization and institutional economics to understanding agri-food chain performance are offered. Implied performance measures that emerge from the explanatory models are discussed and the paper concludes with research challenges still needing to be addressed regarding the measurement of performance in agri-food chains.

THE VERTICAL COORDINATION CONTINUUM

Based on the work of Williamson (1973; 1975), Mahoney (1992) and Milgrom and Roberts (1992), Peterson et al. (2001) posited that the vertical coordination continuum has five major categories of vertical coordination strategy: spot markets, specification contracts, relation-based alliances, equity-based alliances, and vertical integration. Figure 1 contains a table of the relevant definitions for each category of vertical coordination strategy. The latent variable linking the five categories into a true continuum is the intensity of control that the alternative strategies employ to

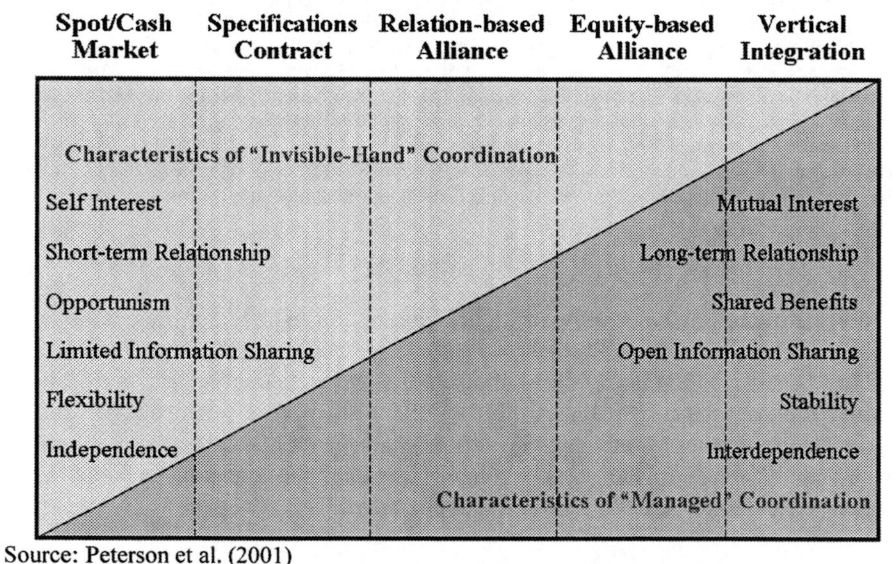

Source: Peterson et al. (2001)

Figure 1. *Strategic options for vertical coordination*

assure that proper coordination occurs (i.e., coordination with minimum potential for error). Coordination strategies move from low levels of *ex ante* control intensity (spot markets) to high levels of *ex post* coordination control (vertical integration) while passing through several transitional levels of ever-increasing intensity (specification contracts, relations-based alliances and equity-based alliances).

THE PWH MODEL OF VERTICAL COORDINATION STRATEGY SELECTION

The main objective of the PWH theoretical framework is to identify the decision-making process where decision makers accommodate issues of asset specificity, complementarity and coordination strategy feasibility in their coordination strategy choices. Peterson, Wysocki and Harsh (Wysocki et al. 2003) modelled a firm's decision about which strategy to pursue on the continuum as a four-step decision process. Figure 2 presents this framework. The framework is based on the presumption that a firm already exists and by intention or habit has already established a position on the continuum. The first decision step involves a process-initiation question: Is the perceived cost of the current coordination strategy too high relative to an available alternative strategy (Node1, Figure 2)? An existing strategy may be too costly for one of two reasons: (1) coordination errors regularly expose the firm to the opportunism of trading partners or result in chronic over- or under-production versus demand, or (2) the strategy is more costly to execute than the coordination errors it is designed to control.

If a firm decides it is dissatisfied with the current strategy from a costliness viewpoint, the second critical question becomes: Would an alternative strategy reduce the perceived costliness of coordination (Node 2, Figure 2)? The answer to this question depends upon whether or not another strategy would better match the intensity (and cost) of coordination control with the costliness of coordination errors. The match is judged better or worse under the logical principle that the more costly the errors, the more intense the control needed and conversely, the less costly the errors the less intense the control.

Again, drawing upon Williamson (1973; 1975), Mahoney (1992) and Milgrom and Roberts (1992), Peterson, Wysocki and Harsh (Wysocki et al. 2003) identify two criteria that can be used to assess the costliness of a coordination error for a given transaction: (1) asset specificity[1], and (2) complementarity[2]. The costliness of a coordination error thus rises with both the level of asset specificity and the level of complementarity. Managers need to assess both of these variables relative to specific transactions and then select a coordination strategy that matches the intensity of control with the costliness of a coordination error. If there is no better match, the perceived costliness diminishes or becomes less important to the decision maker.

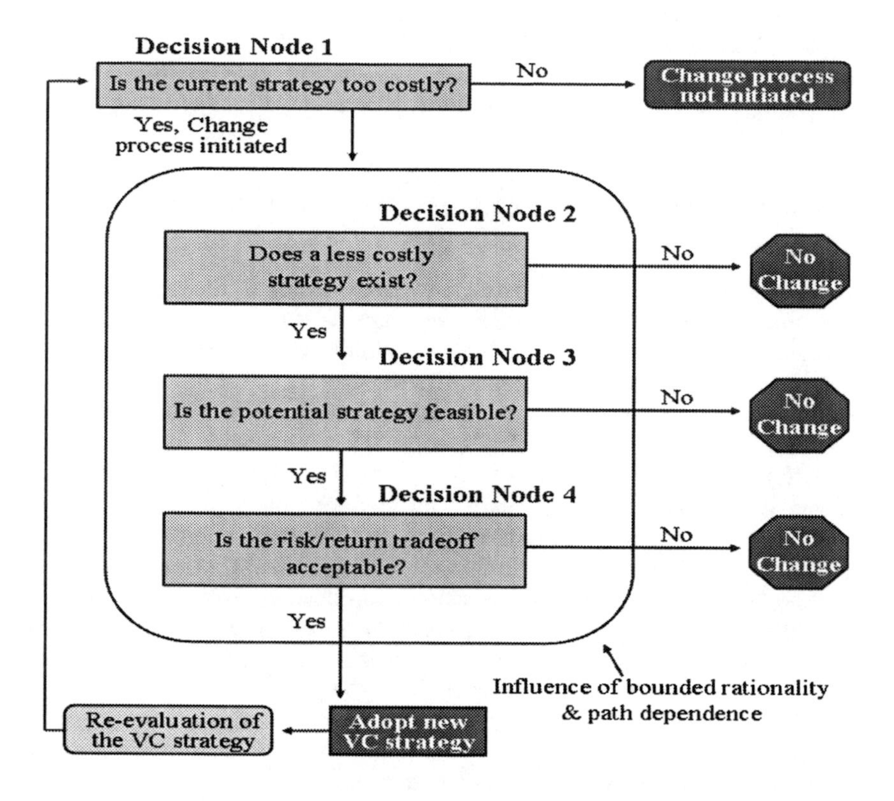

Figure 2. *PWH model of vertical coordination strategy selection*

The third relevant question[3] becomes: Is the potential alternative feasible (Node 3, Figure 2)? Feasibility can be conceived as arising from four conditions: (1) capital availability (does the decision maker have the capital required to implement the strategy?), (2) existence of compatible partners (does the decision maker have a transacting partner who will meet the needs of the strategy being implemented?), (3) control competence (given that each coordination strategy has a different intensity of control, decision makers must examine their competence in exercising the type of control required by the strategy to be implemented; willingness as well as skill is key to competence), and (4) institutional acceptability (an obvious test of institutional acceptability is whether or not a particular strategy is legal, e.g., not in violation of antitrust laws; institutional acceptability is a broader concept that defines what economic behaviours or strategies are deemed appropriate by given social, cultural, industrial or group norms, the core values of the firm).

A recent USDA ruling illustrates the importance of institutional acceptability and its impact on decision makers in an agri-food chain. Creekstone Farms Premium Beef LLC asked the USDA in February 2004 for permission to test all its cattle for mad-cow disease so the company could restart sales to Japan. The USDA subsequently denied Creekstone's testing request because it would have implied a consumer-safety concern that it said was not supported by scientific evidence (Adamy 2004). The USDA maintained that if some beef is labelled as mad-cow tested, it could suggest that untested beef may not be safe. In this particular situation, Creekstone would be prohibited from changing their vertical coordination strategy because of lack of institutional acceptability. Whether or not a particular alternative strategy is deemed feasible will depend on the decision maker's overall assessment of the above four conditions. Any one condition may create enough concern that a "no" decision about willingness to change will result.

Assuming that an alternative is deemed implementable, the fourth and final question in the change process becomes relevant: Does the alternative provide a risk/return trade-off acceptable to the decision maker (Node 4, Figure 2)? With this fourth question, the explicit task of balancing these potential returns and risks is added to the framework. Obviously, the decision maker's risk preferences will be a critical input to answering this question. Based on the decision maker's risk preference it seems fair to predict that any alternative strategy must meet the test that the perceived risk/return trade-off of the alternative is superior to the current strategy if change is to occur.

The framework of Figure 2 proposes that only a "yes" answer to all four of the relevant strategic questions will result in a changed coordination strategy. A "no" at any point stops the process from starting or continuing. A feedback loop is also presented in the framework to make it clear that the process of coordination strategy evaluation is a dynamic one. As transaction conditions, resource availability and strategy potentials change, the chance to create less costly coordination also changes. As an industry evolves, optimal coordination strategies for individual firms within the industry may move in either direction along the continuum depending upon changes in asset specificity, complementarity, feasibility and risk/return trade-offs.

APPLICATION OF THE PWH MODEL TO AGRI-FOOD SITUATIONS

The PWH model of vertical coordination strategy selection has been tested on decision makers in the Michigan celery and public-variety field seed agri-food chains, and in Brazil, at the São Paulo fresh-produce market.

Application of the PWH model to the Michigan celery and public-variety field seed agri-food chains

A qualitative and quantitative approach was undertaken to test the PWH model in the Michigan celery and public-variety field seed agri-food chains. Structured, in-depth, face-to-face interviews were conducted with 25 decision makers in these agri-food chains. Interview transcripts were coded and categorized as responses to each of the four decision-node questions. Qualitative and quantitative analyses were used to test four research propositions (Wysocki et al. 2003):

- RP_1: IF a decision maker is willing to change vertical coordination strategy THEN a "yes" assessment has been made at ALL decision nodes (necessary conditions for strategy change).
- RP_2: IF a "yes" assessment is made at ALL decision nodes, THEN a decision maker is willing to change vertical coordination strategy (sufficient conditions for strategy change).
- RP_3: IF a decision maker is not willing to change vertical coordination strategy THEN a "no" assessment has been made at one OR more decision nodes (necessary conditions for status quo).
- RP_4: IF a "no" assessment is made at ANY one decision node, THEN a decision maker is not willing to change vertical coordination strategy (sufficient conditions for status quo).

For research propositions 2, 3 and 4, 100 percent of the cases upheld the PWH model. Testing of research proposition one revealed that 10 out of the 13 cases studied upheld the PWH model. Three decision makers indicated that they were "unsure" about the risk return trade-off (Decision Node 4 in Figure 2), leading to a classification of the research proposition not holding. Furthermore, discriminant analysis revealed a high degree of predictability with the PWH model. The results of the discriminant analysis of the interview responses revealed that the ability of an alternative to reduce uncertainty was critical to the willingness and unwillingness to change strategies. The acceptability of the risk/return trade-off, the final decision variable, was as important as the reduced costliness of a coordination error in the analysis. Implementability was found qualitatively and quantitatively significant, although not to the same extent as costliness of a coordination error or acceptability of the risk/return trade-off. There was strong quantitative and qualitative evidence to support research propositions 2, 3 and 4 and moderately strong evidence to support research proposition 1. Even with the relatively imprecise instruments (i.e., the decision variables), it appears the proposed framework was a reasonable model of how decision makers view the decision process for altering a vertical coordination strategy.

Application of the PWH model to the São Paulo fresh-produce market in Brazil

Manville and Peterson (in prep.) provided another empirical test for the PWH model for vertical coordination strategy selection, drawing evidence from four firms' procurement strategies in São Paulo's fresh-produce markets. The evolution of firms' coordination strategy decisions for fresh-produce procurement were analysed through seven cases using the PWH framework.

Case analysis and application to the PWH model led to several insights. In six of the seven cases (in one case a firm that began with a vertical integration strategy and subsequently de-integrated) firms consistently tended to shift their strategies toward ones offering greater intensities of control. In general, this does not necessarily indicate that the firms initially made the 'wrong' coordination strategy choice. Instead, there are suggestions that both retailers and suppliers in this produce market benefited from the gradual intensification of coordination control, over which period they learned to work better with one another and adjusted to the greater levels of interaction and interdependence, while incompatible partners were identified and weeded out. There were also important driving forces, namely the shifts in underlying demand and supply conditions faced by the firm that also drove these strategic shifts towards ever-increasing levels of control (Manville and Peterson in prep.).

Second, differentiation between issues of complementarity and asset specificity permit a considerably richer insight into issues of coordination than a single-minded focus on asset specificity. The defining difference between the two concepts highlights that coordination can be costly even where there is no reason to anticipate opportunistic behaviour on the part of either party, i.e., in situations of complementarity where both parties will benefit from the successful completion of transactions (Manville and Peterson in prep.).

Third, it seems that as one moves from lesser to greater degrees of control intensity, one can observe early shifts rightward to be oriented to resolving issues of complementarity, with the resolution of these leading to gradual increases in the levels of asset specificity implicit in the transaction; and this in turn leading to the need for additional shifts in coordination strategy. For example, in the first phase of one firm's coordination strategy, they sought to facilitate the achievement of complementarity benefits by tightening control through a supplier registry, which led to relationship-specific investments whose asset specificity later needed to be accommodated through further shifts in coordination.

In this situation, case-study analyses provided significant support for the PWH model, as well as new insight into the firms' coordination strategy decisions that might be forgone when analysed using a model that lacks the operational approach and consideration of incentives and feasibility guiding the PWH model. The results support the hypothesis that the PWH framework provides empirical and theoretical insight into managers' coordination decisions. Analyses of the evolution of the firms' coordination strategies appear to provide powerful evidence supporting the determinants and processes of decision-making that Peterson, Wysocki and Harsh posit in their model. The case-study results provide considerable insight into issues of incentives and feasibility, as well as the influence that asset specificity, complementarity and risk/return criteria have on the coordination strategy decision.

AGRI-FOOD CHAIN FORMS: UNINTEGRATED AND INTEGRATED CHAINS

A supply chain, whether in the agri-food system or not, is "an association of customers and suppliers who, working together yet in their own best interests, buy, convert, distribute, and sell goods and services among themselves resulting in the creation of a specific end product" ((National Research Council 2000, p. 22). By this definition, every firm is part of a supply chain, and supply chains have always existed even within the context of spot-market interfaces between firms, level by level, in a vertical chain. Traditional supply chains coordinated by open markets can be referred to as unintegrated. Today, integrated supply chains are moving beyond open-market association to "an association of customers and suppliers who, using management techniques, work together to optimize their collective performance in the creation, distribution, and support of an end product" (National Research Council 2000, p. 27). An integrated supply chain can only be optimized when the chain participants function together to improve the performance of the whole chain (Peterson 2002).

The benefits of an integrated supply chain are argued to be many: inventory reduction throughout the chain; reductions in supplier redundancy; reduced transaction costs, frictions and barriers; increased functional and procedural synergies between chain members; faster response to changing market demands; lower operating and investment costs across the chain; and shorter product realization cycles and lower product development costs (National Research Council 2000). These benefits mirror quite well the ones expected from the concepts of transaction-cost economics (Williamson), the theoretical foundation for much of supply-chain management. As information is shared, functions rationalized and system flows made more certain across the supply chain, transaction costs fall and the responsiveness and profitability of the whole chain rises.

An earlier section of the paper described the usefulness of applying the PWH model to the Michigan celery and public-variety field seed agri-food chains. In fact, 9 out of the 25 firms studied, were operating in unintegrated agri-food chains using a spot market strategy as their primary market-coordinating mechanism. The remaining 16 firms were operating in integrated agri-food chains by utilizing contracts, relation-based alliances, equity-based alliances and vertical integration (Wysocki 1998).

LEARNING SUPPLY CHAINS

Learning supply chains are formed from the union of learning organizations and integrated supply chains. Learning organizations are those organizations that have developed knowledge management systems that allow them to adapt continuously to their environment based on learning (Senge 2000). Learning is "the process by which knowledge assets are increased over time" (Seemann et al. 2000, p. 91). Learning organizations have "explicit management efforts to build intellectual capital in support of the firm's strategy" (Seemann et al. 2000, p. 92). Learning arises from the creation and sharing of both tacit and explicit knowledge.

For our purposes, a learning supply chain can be defined as an integrated supply chain that has an added dynamic, agile ability to learn from and respond to changing market environments. The added capacity of a learning supply chain is the knowledge or intellectual capital held and applied collectively by the supply chain (Peterson 2002). Mason-Jones and Towill (1999) argue that the key added capacity formed by a learning supply chain is the simultaneous feeding of end-user market information to all supply-chain participants. Dyer and Nobeoka (2000) describe the crucial addition to capacity as highly effective, free-flowing interfirm knowledge transfers. For De Vries and Brijder (2000), the two-way knowledge sharing of upstream functional/operational knowledge and downstream contextual (market) knowledge between chain members is the significant added capacity.

ALTERNATIVE MODELS FOR LEARNING SUPPLY CHAINS

Rice and Hoppe (2001) suggest three different models for future supply chain versus supply-chain competition. Each of these models has implications for agri-food chain performance and for distribution of costs and benefits generated by each. The three models are the channel master, the chain web and the chain organism.

The channel-master model

Inter-chain competition in the channel-master model centres on a dominant firm, or channel master, in each supply chain. The dominant firm specifies the terms of trade across the entire supply chain, and the supply chain rises or falls competitively based on the coordinating skill of this firm. Most existing automotive supply chains (except one noted shortly) are examples of this model. The model is also well represented in the agri-food system. It may even be the prevailing model today given the pervasiveness of 'integrators' in a number of agri-food chain settings. Tyson, Purdue and Smithfield would all represent examples (Peterson 2002).

The chain-web model

The chain-web model is an integrated supply chain in which individual firms compete with others outside their respective supply chains but based on their own supply network capabilities. Individual firms may well be members of multiple supply chains. They will also connect and disconnect from these chains as they find it in their own best interest to do so. Integration is created through a broad range of relationships, including joint ventures, joint marketing arrangements, collaborative initiatives in systems and processes, etc. (Peterson 2002). The picture that emerges for this model is a web of interfirm relationships that is continually changing shape, dimension and membership. The driving logic of the web remains the individual strategies of the firms. The computer industry is largely characterized by this form of chain organization (Rice Jr. and Hoppe 2001). In many respects, smaller processing and food-manufacturing firms must already behave in this manner in order to serve food retailers. The adoption of category management creates a significant incentive for these smaller firms to combine and recombine in order to maintain access to the retailer (Peterson 2002).

The chain organism

The chain-organism model is an integrated supply-chain model in which the supply chain itself competes as one entity, one dynamic organism. To distinguish this model from the channel master, note that there cannot be a dominant firm in the chain. All firms share the decision making and find themselves inherently interdependent in their ability to act. The strength of this interdependence is central to resolving the dilemmas inherent in a learning supply chain (Peterson 2002).

Dyer and Nobeoka (2000) present extensive case evidence that Toyota is a high-performance knowledge-sharing network, that is, a learning supply chain. Their evidence also suggests that Toyota is not a channel master; rather it has created a chain organism. Their case is made through a detailed examination of the various coordination mechanisms used by Toyota to establish and maintain the character of the supply-chain network, most especially its focus on knowledge management. Network-level knowledge management processes include a supplier association, Toyota's operations management consulting division, voluntary small-group learning teams, and interfirm employee transfers. Collectively these processes create a network that promotes the sharing of knowledge, both explicit and tacit, in multilateral and bilateral settings (Peterson 2002).

AGRI-FOOD CHAIN PERFORMANCE MEASURES

Research on supply chains has traditionally focused on identifying performance measures for specific firms in a given supply chain. In a world where agri-food supply chains increasingly compete against other supply chains, performance measures need to include the entire supply chain, not just individual firms within a given supply chain. This section outlines possible performance measures for agri-food chains including chain efficiency, profitability, distribution of returns, chain responsiveness, prices, and the value that is delivered.

Increased chain efficiency

While increased efficiency can be measured in numerous ways, it would seem reasonable to consider a limited number of metrics, as agri-food chain efficiency measures of performance. Transaction-cost metrics could include the percent of products sold and introduced the same year, the percent of product sales introduced the same year, the cycle time needed to develop and deliver a new product, inventory reduction and supplier redundancy reduction, and lower operating and investment cost.

Shorter product realization cycles could be measured and metrics could include: inventory value, inventory turns (annual cost of sales divided by annual average inventory value), return on sales, cash-to-cash cycle time (this is the time it takes from when a company pays its suppliers for materials to when it gets paid by its customers), activity cycle time (amount of time it takes to perform a supply-chain activity such as order fulfilment, product design, assembly which can be measured within a firm or more importantly across an entire chain), upside flexibility (the ability of a chain to respond quickly to additional order volumes, which could be measured as the percentage increase over the expected demand for a product that can be accommodated), and outside flexibility (the ability to provide the customer quickly with additional products outside the bundle of products normally provided).

Increased profitability, distribution of returns, and equity versus equality

Measuring profitability for an entire supply chain is no simple task. One immediate problem in today's business environment is obtaining accurate financial information for privately held firms. It is possible to obtain financial data from publicly-held entities, but financial information on privately held firms is usually a guess at best.

How are profits distributed across the agri-food chain? The answer to this question may depend, in part, on the supply-chain form. One would not expect to see the distribution of returns to be the same for a channel-master supply chain as opposed to a chain web or chain organism. The use of institutional economics could be very helpful in answering questions of equity and equality when it comes to agri-food supply-chain performance measures.

Chain responsiveness

Chain responsiveness could become increasingly important in the future. Customers continue to gain market power and knowledge by ready access to information – virtually wherever, whenever and however they want it. Retailers must provide value propositions and shopping experiences that keep customers coming back, even in a world of total information transparency. The world's top retailers are rapidly expanding across geographies, channel formats and product/service categories, blurring market segments and devouring market share. Competitors must successfully differentiate themselves in order to survive (IBM Business Consulting Services 2004).

The IBM Business Consulting Services (2004) calls this focus on the shopper 'customer centricity'. Greater customer focus must go beyond the superficial by addressing all the basic building blocks of the organization. The status quo must change from disconnected, multiple channels and silos to a unified orchestration of the customer experience. Retailers will need to be capable of delivering a unified seamless customer experience that treats customers as the unique individuals they are.

So the important question might become – is the agri-food supply chain able to meet the expectations of its customers? Metrics that could be measured include: order fill rates, on-time delivery rate, value of backorders, number of backorders, frequency and duration of backorders, line-item return rate, quoted customer response time and on-time completion rate, value of late orders and number of late orders, frequency and duration of late orders, and number of warranty returns and repairs. From an end-user's perspective are prices being driven lower over time and is value or the relationship of the bundle of services to price improving over time?

IMPLIED PERFORMANCE MEASURES ARISING OUT OF A MULTI-DISCIPLINARY APPROACH TO UNDERSTANDING AGRI-FOOD CHAIN PERFORMANCE

A deeper understanding of agri-food chain performance is likely to require a multi-disciplinary approach. This multi-disciplinary approach could include insights from transaction-cost economics, industrial organization, strategic management and institutional economics. This section concludes with a discussion of the implied agri-food chain performance measures that result from the incorporation of the PWH model, learning supply-chain governance structures, and application of additional economic theories.

Transaction-cost economics

Much has been written and studied about transaction-cost economics and the work of Coase, Williamson and others. Application of concepts such as asset specificity, quasi-rents, principal-agent problems and opportunism help to explain agri-food chain relationships and their impacts on chain performance. For example, it is widely known that entering the Wal-Mart supply chain as a supplier will require the use of specific assets such as computer and data transmission systems that are required to interact with Wal-Mart's just-in-time purchasing and inventory systems. The sheer buying power of a company like Wal-Mart causes agri-food suppliers to worry about opportunistic behaviour on the part of Wal-Mart. This is especially true when suppliers find themselves in a position of having a significant portion of their sales devoted to one customer.

Industrial organization and strategic management theory

Just as with transaction-cost economics, an extensive body of knowledge exists within industrial organization and strategic management theory to help explain agri-food chain relationships and their impacts on chain performance. The structure–conduct–performance paradigm is one of the cornerstones of industrial-organization theory. The basic tenet of the S–C–P paradigm is that the economic performance of an industry is a function of the conduct of buyers and sellers, which, in turn, is a function of the industry's structure (Mason 1939; Bain 1956). Economic performance is measured in terms of welfare maximization (resources employed where they yield the highest valued output). Conduct refers to the activities of the industry's (for our purpose, agri-food chain's) buyers and sellers. Sellers' activities include installation and utilization of capacity, promotional and pricing policies, research and development, and interfirm competition or cooperation. Industry structure (the determinant of conduct) includes such variables as the number and size of buyers and sellers, technology, the degree of product differentiation, the extent of vertical integration, and the level of barriers to entry (Scherer 1980, p. 4).

For example, the meat-processing industry continues to become more concentrated, with two or three firms capturing the majority of the market share (structure). These firms are increasingly able to impose their will on suppliers (e.g., producers) to provide specific products and services they specify. Often, this is done through contracting (conduct). Producers in these concentrated industries often have few alternatives for their product or services, which can lead to revenues that are regulated by the few, large, buyers (performance).

Institutional economics

Institutional economics incorporates a theory of institutions into economics. It builds on, modifies and extends neoclassical theory. It retains and expands on the fundamental assumption of scarcity and hence competition – the basis of the choice-theoretic approach that underlies microeconomics. It has developed as a movement within the social sciences, especially economics and political science, that unites theoretical and empirical research examining the role of institutions in furthering or preventing economic growth. It includes work in transaction costs, political economy, property rights, hierarchy and organization, and public choice. Most scholars view the work of Ronald Coase as a central inspiration for the field (North 1992).

Social capital, trust and property rights are additional tenants of institutional economics that are relevant for a better understanding of agri-food chains. Social capital is a feature of social organizations, such as trust, norms and networks, that can improve the efficiency of society by facilitating coordinated actions (Putnam 1993). Social capital is often inherent to the structure of relations between and among actors. Each variety of social capital consists of some aspect of social structure, and each facilitates certain actions of actors – persons or corporate actors – within the structure (Coleman 1988). The economic property rights of an individual over a commodity or an asset are the individual's ability, in expected terms, to consume the good or the services of the asset directly or to consume it indirectly through exchange. These can include (1) the right to use an asset, (2) the right to earn income from an asset and contract over the terms with other individuals, and (3) the right to transfer ownership rights permanently to another party. Legal property rights are the property rights that are recognized and enforced by the government (Barzel 1997; Eggertsson 1990).

Consider the institutional economics aspects of the highly concentrated meat-processing industry. An institutional economics perspective would explore in detail, the property rights of the parties involved and how this impacts the distribution of a performance measure like profit. For example, meat processors like Tyson specify the genetics and husbandry practices of their broiler producers. Tyson has the property rights to the genetics and the power to enforce the husbandry practices. One might also argue that Tyson is taking on more risk in this relationship relative to producers, and as a result is entitled to more of the profit generated from this integrated agri-food supply chain.

Implied agri-food chain performance measures as they relate to PWH model, learning supply-chain governance structures and application of additional economic theories

The PWH model draws heavily on transaction-cost economics. The implied performance measures include the costliness of coordination errors, both in magnitude and probability, and the costliness of transaction within a given governance structure. The trade-off between costliness of coordination errors and costliness of transactions contributes to overall efficiency of the supply chain. This efficiency can be measured as technical efficiency, transaction efficiency, innovation ability, profitability (across firms and entire supply chains), responsiveness and equity.

While the PWH model is useful for understanding overall efficiency measures, the learning supply-chain framework is well suited for understanding relative efficiency of the three types of integrated supply chains versus unintegrated supply chains. Relative efficiency measures include technical efficiency, transaction efficiency, innovation ability, profitability (across firms and entire supply chains), responsiveness and equity.

There is a connection between the intensity of coordination control and supply-chain governance structure. The intensity of coordination control (from low to high across the vertical coordination continuum) is the underlying variable affecting coordination choice in both unintegrated and integrated supply chains. It has already been shown how PWH analysis can be applied to firms in an unintegrated and integrated chain to predict changing relationships between stages. With learning supply chains, an important performance issue becomes one of 'who' controls the intensity of coordination control. In the channel-captain model, the channel captain controls. Supply chains organized as chain webs rely on individual networks within the chain for control. The chain collective of the chain organism is responsible for regulating the intensity of control.

The coordinating skills of the channel captain dictate the success or failure of the chain. For example, if largely based on specifications contracts, the chain may not keep pace with changing marketplace conditions. If the channel master wants to organize in a way that puts individual supply-chain members at sub-optimal levels, the performance of the entire chain will be sub-optimal.

Under the chain web form, organizations that connect or disconnect for self-interest may not be in the best interest of the entire web chain. The motivation to share valuable individual information may be less present in a chain-web form. Continual reconfiguration of the chain web maximizes the potential loss of knowledge, which can lead to increased transaction costs and missed market opportunities.

Substantial barriers exist regarding the formation of learning supply chains and chain organisms in the agri-food system. To date, the authors are hard-pressed to describe an example of a chain organism in the agri-food system. The channel-master and chain-web forms may function quite well to meet needs of agri-food end-users even if the added benefits of knowledge management are absent. The bottom line is: would agri-food chain performance be improved under the chain-organism model? The answer remains unclear.

There is redundancy, overlap and inter-connectedness to measuring agri-food chain performance when using the PWH model, learning supply-chain governance structures, and application of additional economic theories. For example, coordination error could be a failure of responsiveness or transaction inefficiency.

CONCLUDING REMARKS

A number of challenges remain for researchers wishing to measure performance in agri-food supply chains. Access to information issues exist. Under increasingly closed chains, how does one gain access to the needed information? Measurement issues are common as well. How does one measure profitability across a supply chain? Should variables be measured in the same place across the entire chain? Even if one were able to measure the performance variables stated earlier, would that tell enough of the story? The rich history of institutional economics tells us it is still useful to ask the distributional issues of who wins and loses. Even if one were able quantify answers to the PWH key variables across an entire chain, would we learn that much more? As with many of the performance issues, the PWH model is complex enough to apply to a single firm in a supply chain, let alone to an entire agri-food chain.

In the end, understanding agri-food chains is likely to require a multidisciplinary approach. Measuring agri-food chain performance is a complex task that requires innovative techniques. The PWH, learning supply-chain governance structures, and application of additional-economic-theories model can be useful in understanding and measuring performance in agri-food chains.

NOTES

[1] Asset specificity is the degree to which an asset can be redeployed to alternative uses and by alternative users without sacrifice of productive value.

[2] Complementarity exists when the combining of individual activities across a transaction interface yields an output larger than the sum of outputs generated by individual activities.

[3] The PWH model originally included programmability as a third decision (addressed before the feasibility question) in their model. This step is omitted, as it was not found to be significant in our initial work. It is expected that firms will only seriously consider coordination strategies that they consider to be programmable, so that programmability is addressed implicitly in the range of alternatives that the firm initially chooses to consider.

REFERENCES

Adamy, J., 2004. U.S. rejects meatpacker's bid to conduct mad-cow testing. *Wall Street Journal,* 243 (71), B6.

Bain, J S., 1956. *Barriers to new competitions*. Harvard University Press, Cambridge.

Barzel, Y., 1997. *Economic analysis of property rights*. 2nd edn. Cambridge University Press, Cambridge.

Coleman, J.S., 1988. Social capital in the creation of human capital. *American Journal of Sociology,* 94 (Suppl.), 95-120.

De Vries, E. and Brijder, H.G., 2000. Knowledge management in hybrid supply channels: a case study. *International Journal of Technology Management,* 20 (5/6/7/8), 569-587.

Dyer, J.H. and Nobeoka, K., 2000. Creating and managing a high-performance knowledge-sharing network: the Toyota case. *Strategic Management Journal,* 21, 345-367.

Eggertsson, T., 1990. *Economic behavior and institutions*. Cambridge University Press, Cambridge.

IBM Business Consulting Services, 2004. *The retail divide: leadership in a world of extremes*. IBM Global Services, Somers.

Mahoney, J.T., 1992. The choice of organizational form: vertical financial ownership versus other methods of vertical integration. *Strategic Management Journal,* 13 (8), 559-584.

Manville, D.Y. and Peterson, H.C., in prep. Coordination strategy decisions in São Paulo's fresh produce markets: an empirical test of the Peterson, Wysocki & Harsh Framework. *International Food and Agribusiness Management Review*.

Mason-Jones, R. and Towill, D.R., 1999. Total cycle time compression and the agile supply chain. *International Journal of Production Economics*, 62 (1/2), 61-73.

Mason, E.S., 1939. Price and production policies of large-scale enterprise. *American Economic Review*, 29, 61-74.

Milgrom, P. and Roberts, J., 1992. *Economics, organization and management*. Prentice-Hall, Englewood Cliffs.

National Research Council, Committee on Supply Chain Integration, 2000. *Surviving supply chain integration: strategies for small manufacturers*. National Academy Press, Washington. [http://www.nap.edu/catalog/6369.html]

North, D.C., 1992. *The new institutional economics and development*. Washington University, St. Louis. [http://www.econ.iastate.edu/tesfatsi/NewInstE.North.pdf]

Peterson, H.C., 2002. The "Learning" supply chain: pipeline or pipedream? *American Journal of Agricultural Economics*, 84 (5), 1329-1336.

Peterson, H.C., Wysocki, A.F. and Harsh, S.B., 2001. Strategic choice along the vertical coordination continuum. *International Food and Agribusiness Management Review*, 4, 149-166. [http://www.ifama.org/nonmember/OpenIFAMR/Articles/v4i2/149-166.pdf]

Putnam, R.D., 1993. *Making democracy work: civic traditions in modern Italy*. Princeton University Press, Princeton.

Rice Jr., L.B. and Hoppe, R.M., 2001. Supply chain versus supply chain: the hype and the reality. *Supply Chain Management Review*, 79 (9/10), 46-55.

Scherer, F.M., 1980. *Industrial market structure and economic performance*. Houghton Mifflin, Boston.

Seemann, P., DeLong, D., Stukey, S., et al. 2000. Building intangible assets: a strategic framework for investing in intellectual capital. *In:* Morey, D., Maybury, M. and Thuraisingham, B. eds. *Knowledge management: classic and contemporary works*. MIT Press, Cambridge, 84-98.

Senge, P., 2000. The leader's new work: building learning organizations. *In:* Morey, D., Maybury, M. and Thuraisingham, B. eds. *Knowledge management: classic and contemporary works*. MIT Press, Cambridge, 19-52.

Williamson, O.E., 1973. Markets and hierarchies: some elementary considerations. *American Economic Review*, 63 (2), 316-325.

Williamson, O.E., 1975. *Markets and hierarchies: analysis and antitrust implications: a study in the economics of internal organization*. Free Press, New York.

Wysocki, A.F., 1998. *Determinants of firm-level coordination strategy in a changing agri-food system*. Michigan State University, East Lansing. Ph.D. Thesis, Michigan State University Department of Agricultural Economics

Wysocki, A.F., Peterson, H.C. and Harsh, S.B., 2003. Quantifying strategic choice along the vertical coordination continuum. *International Food and Agribusiness Management Review*, 6 (3), 1-15.

CHAPTER 14

FROM CORPORATE SOCIAL RESPONSIBILITY TO CHAIN SOCIAL RESPONSIBILITY

Consequences for chain organization

JACOBUS J. DE VLIEGER

LEI, Wageningen University and Research Centre, P.O. Box 29703, 2502 LS The Hague, The Netherlands. E-mail: Koos.devlieger@wur.nl

Abstract. In both the political and the scientific field the ideas about corporate social responsibility (CSR) have changed over the last 10 years. These changes are more or less connected with the stages in the environmental policy of firms as De Ron et al. (2002) describe them – put in order, administer and integrate – and with the development of political ideals about CSR. The relation between the social responsibilities of firms and chain organization is founded on the 'credence' characteristics of social responsibility. These credence characteristics make market information necessary to ensure market information is symmetric in every stage of the production chain; otherwise market failure is unavoidable and inferior products will drive out good-quality products. The incorporation of transparency on CSR into a chain is possible in a product-oriented (LCA) or a company-oriented way (company-certifying systems).

The integration of CSR in chains is accompanied by management efforts, both within and without the company. Firms must implement the strategy in their whole enterprise, from top management down to shop-floor workers. This is often done with the help of existing quality management concepts such as INK. In extra-company management (chain cooperation) both the content (what) and process (how) are important.

The last part of the paper covers in more detail how to manage the chain with respect to CSR. In this part we will discuss competence analysis and monitoring with the help of the model of Doz and Hamel (1998). To transform this static model into a dynamic one we propose using the model of Ring and Van der Ven (1994) for the cyclical development of cooperation and the model of Lewicki and Bunker (1996) for the build-up and development of trust.

To unite these more general models with the more specific approach in the field of CSR we integrated cooperation with the recently developed European Corporate Sustainability Framework. Effectively, we chose a company approach and the integration of CSR with quality-assurance models, because this approach is a promising way to start CSR in chains and manage it.

Keywords: corporate social responsibility; chains; networks; management.

INTRODUCTION

In 1987 the Brundtland Report underlined the importance of sustainability and called for "a form of sustainable development which meets the needs of the present

C.J.M. Ondersteijn et al. (eds.), Quantifying the agri-food supply chain, 191-206.

without compromising the ability of future generations to meet their own needs". The key issue in the Brundlandt Declaration is: "*A better life for everyone without destroying our natural resources for future generations*". In the years that followed the declaration, government and firms have worked on its translation into laws, rules and codes of firm conduct. This has resulted in a number of developments, in practice as well as in theory.

On the political level the United Nations Conference on Environment and Development (UNCED, Rio de Janeiro 1992) was held. The conclusion of this conference was that actions are necessary in the field of poverty, modification of non-sustainable production and consumption patterns and the conservation and management of natural sources. The World Summit on Sustainable Development (WSSD, Johannesburg 2002) acknowledged the importance of sustainable development and underlined that the input of all societal groups is important and cannot be missed. Therefore, the United Nations formulated 9 principles (with respect to human rights, labour and environment) and asked firms to apply these principles (www.unglobalcompact.org). The OECD (2001) guidelines for multinationals are considered to be the standard for the behaviour of firms in international trade and a basis for the formulation of an internal code of conduct for corporate firms.

In the years since the Brundtland Declaration the content of sustainability has been broadened from Planet to include Planet, People and Profit. In The Netherlands this was reflected in 'De Winst van Waarden' (The Profit of Values), a policy advice from the Social-Economic Council (SER 2000). In this advice, the Council distinguishes two important elements for Corporate Social Responsibility (CSR), namely:

1. Aiming firm activities at a contribution to the welfare of the whole society in the long term, based on the dimensions Profit, People and Planet (content);
2. Having a relationship with stakeholders on the basis of transparency and dialogue, and providing answers to the legitimate questions of society (process).

At the same time the view on the role of government and policy became clearer. The Dutch government and the Dutch Ministry of Agriculture, Nature and Food Quality (LNV) believes that CSR underlines the active and voluntarily accepted societal role a firm fulfils – but not without obligations – and that CSR goes beyond simply applying the law and its rules. This viewpoint means the Dutch government sees CSR in the first place as the responsibility of the firms themselves, whilst at the same time underlining that great political interests are involved. That is the reason why the government thinks stakeholder dialogue is of importance. This government viewpoint is also in line with that of the EU Commission, as presented in the white paper on CSR (Commission of the European Communities 2002).

Worldwide more and more corporate firms are involved in sustainability initiatives and report their activities in annual reports. The Global Reporting Initiative (GRI) has developed guidelines for reporting on sustainability (GRI 2002). For instance, a report must be complete and comparable. The guidelines also mention 5 components that have to be present in the report:

- vision and strategy of the organization;
- organization profile;
- governance structure and management systems;
- a GRI content index (a table identifying where the information requested by the guidelines is located in the report); and
- performance indicators.

In the theoretical field of sustainability and CSR, Wood (1991) made an important contribution with her so-called Corporate Social Performance model, in which she distinguishes principles, processes and performances. With respect to principles, Wood made clear there are different levels:

- The institutional principle relates to legitimacy granted by society. Legitimacy relates the responsibility of firms as a societal institution.
- The organizational principle is related to public responsibility – a relational aspect between the firm and its own surroundings.
- The individual principle is based on the idea that a manager is a moral actor.

These principles, however, are neither universal nor absolute, but connected to time and culture.

Wood (1991) differentiates the processes into context, stakeholder management and the management of changes, based on the basic strategy of the firm with respect to CSR. Wood uses the continuum of firm behaviour developed by Carroll (1979), who distinguished between a reactive, defensive, adaptive and pro-active strategy. A pro-active strategy means the firm continually monitors and evaluates its context, pays attention to the demands stakeholders make, and develops plans and policies to adapt to changing conditions.

In the results/performances, Wood (1991) distinguished societal effects of the behaviour of the firm (community investment, net profit), the programmes used by the firm to realize CSR (stakeholder dialogue), and the policies developed to handle societal issues and stakeholders' interests (mission and code of behaviour).

The Social-Economic Council of The Netherlands (SER 2000) considered CSR as the nucleus of entrepreneurship. This implies that it must be embedded in the whole firm, from the shop floor to the CEO and vice versa. This is why the management level – a staff or line function – responsible for CSR is important.

Social responsibility goes beyond simply applying law and its rules. The result may be totally different firm policy goals. Further, De Ron et al. (2002) distinguish three stages in the environmental policy of firms: put in order, administer and integrate. During the first phase, firms are forced to comply with the more severe rules of the government, i.e., compliance is the main goal. In the next phase, administer, the firm notices the possibility to gain economic profit from sustainability, so-called eco-efficiency. In this phase the firm goes beyond compliance: the 3Ps are synonymous with 'pollution prevention pays'. In the integration phase the aspects sustainability and strategy are connected and the firm sees sustainability as Profit, People and Planet. De Ron et al. (2002) translated this model of environmental behaviour of firms also into the broad approach of sustainability. For this, in Chain Social Responsibility the goals and vision of the participating firms on social responsibility have to be in line with each other.

In the political, corporate and theoretical developments so far, the focus point has been the role of the single firm, although in a number of studies connections are also made to suppliers and buyers. In this paper the chain is our focal point because of the characteristics of corporate social responsibility, which will be discussed first. In the second part, the influence of these characteristics on the incorporation of CSR in networks and chains is discussed. After, we will discuss a method to analyse the status of CSR incorporation in networks and chains and how to accommodate this process. Finally, some conclusions are given. Both the second and third part of this paper are almost entirely based on recent work within the Agricultural Economics Research Institute (LEI). Studies in this area by Wolfert et al. (2003) and Goddijn and Vlieger (2004) represent the state of the art of corporate and chain social responsibility in The Netherlands.

FROM CORPORATE SR TO CHAIN SR

What is the reason for transforming Corporate SR (CSR) into Chain SR? The answer to this question can be found in the characteristics of social responsibility. CSR can be defined as actions taken by a firm that appear to go beyond the interest of the firm and what is required by law and ethics. As we mentioned before, communication about CSR in the broad sense is an important part of this approach. Communication about what companies are doing in the field of CSR is done in several ways. In a recent study by Pannebakker and Boone (2004) we found that most of the firms in The Netherlands communicate about this aspect to society in general and stakeholders in particular via their annual reports, whereas others concentrate on shareholders' value and communicate mainly with them. Some firms show their responsibility by measures for, and communication with, the local community. However, in most cases firms communicate about CSR by means of the characteristics of their products. They inform consumers about the important aspects of their production process, for instance how much energy has been used. The consumer, however, cannot infer these characteristics before or after the purchase. This means that in the marketplace we talk about 'credence[1] characteristics' of a product. To inform consumers about these characteristics and to give them reliable and correct information, a number of measures have to be taken. The reason for this is that if consumers get imperfect information or if market information is asymmetrically available, market failure will be the result, as Akerlof (1970) showed. If quality cannot be spotted, good-quality products cannot get a price premium and accordingly, only inferior products will be offered in the market. However, the assessment of the quality of credence goods involves high costs both before and after consumption (Becker 2000). For credence quality attributes, both private and public measures are needed to let markets function properly and to guarantee the availability and quality of the necessary information (Caswell and E.M. 1996). For firms a cost-efficient assurance of quality is very important. This can be reached by coordinating the assurance schemes in the whole chain. The reason for transforming corporate social responsibility into chain social responsibility is thus connected with the credence characteristics of social responsible products and the flow of information needed to inform buyers about the quality of the products.

The amount of information available on credence characteristics is crucial. Information is usually considered a public good (non-rival and non-excludable) and for that reason undersupplied in the market. If we, as Antle (1999) suggested, consider information as a 'club' good that is non-rival but excludable, then the government has the task to create a legal framework that enables consumers to obtain and use information. The question, then, is which actor in the food chain should offer the required information, by which means and at which costs? Therefore, different types of contracts and quality systems can be used. To make the food network and chain operate efficiently, further research into the contents and impacts of these contracts and systems is needed.

To inform the consumers effectively about credence quality attributes, such as animal welfare and environmental impacts, labelling may be a solution. Labelling implemented by a certifying agent can take many forms. Labels can be detailed, they can also include information about the process and how it differs from other production practices, and labels can also contain information about the certifying agent (Wood Renck 2002).

CSR TRANSPARENCY

The incorporation of transparency on CSR into networks and chains can be either product- or company-oriented (Wolfert et al. 2003). In practice, often a combination of both approaches is found. For a product-oriented approach the credence quality aspects must be unambiguously allocated to individual products. Life Cycle Analysis (LCA) seems an interesting instrument to employ for this purpose, because it considers the whole lifecycle of a product and therefore the whole chain. Commonly, an LCA is used to determine the environmental impact of a product. These impacts are expressed in units of production. An LCA is mainly focused on the environmental aspects of products and hardly linked to economic, social or other aspects. This is because people and profit aspects are often more related to processes than to products. However, in The Netherlands a task force is working on a feasibility study to include this dimension in LCA and life-cycle thinking (Udo de Haes and Sonnemann 2003). A recent product-related development in chains is the growing number of chain information systems used by chain actors to share information. Mostly these information systems are developed to track and trace products. Food safety and the product responsibility of actors are important reasons behind the growing attention to these systems and to tracking and tracing. Combining these chain information systems with LCA techniques potentially plays an important part in making CSR transparent in chains and networks.

The product-oriented approach to transparency can be considered a company-independent method because the individual company's unique efforts are not accounted for. The company-oriented approach is very useful if, as in many agricultural production processes, there are considerable amounts of by-products and externalities. In this situation it is more realistic to start at the company level to realize CSR because then all companies in the chain can guarantee and prove that they produce in a socially responsible way. However, why would companies engage in CSR activity at all? Van der Schans et al. (2003) mentioned three main reasons:

- CSR as an expression of value-driven entrepreneurship (based on personal norms and values);
- CSR based on clearly-understood self-interest (risk and reputation management);
- CSR based on responsibility and dialogue with society (corporate citizenship).

The diversity of motives implies that CSR as a strategic option is not easy to implement if companies are spread over different locations or different levels or form a part of chains and networks. Further, CSR in a chain or network is the highest ambition level in the CSR development model (De Ron et al. 2002). This means companies have to reorganize themselves internally before they can realize CSR goals at chain and network level. In the first phase of this transforming process firms have to comply with the statutory obligations. In the second phase – control – they must be able to show actual results. In the third phase, CSR must be integrated into their corporate and competitive strategy, which implies cooperation with other parties in networks and chains.

REPORTING ABOUT CSR TO THE PUBLIC, STAKEHOLDERS AND SHAREHOLDERS

As mentioned before, at company level CSR is essentially context-dependent, goes beyond legal rules and is voluntary. This results in many different ways to implement SR, monitor performance and report the results. Several initiatives have been undertaken to define and measure performance (see, e.g., GRI 2002; OECD 2001; Meeusen and Ten Pierick 2002). Berenschot Consultancy (2003) has studied the annual reports of several Dutch companies in order to see how far CSR had been incorporated. They distinguish four levels of engagement with CSR, namely:

1. Globally engaged – restricted reporting;
2. Globally engaged – broad reporting – actual activities;
3. Engaged – broad reporting – actual activities – results;
4. Engaged – broad reporting – actual activities – results – external verification.

The reports of most companies fell into the first two categories. Another problem is how to assess the actual performance of a particular company. Should we compare the company with another company that has the same functions or with a company that complies with the same demand, but in another way? Furthermore, we have to remember CSR goes beyond legal rules and is voluntary.

A third aspect is the dialogue with stakeholders on CSR, which, at the level of the individual company, is very time-consuming. In addition the Non-Governmental Organizations (NGOs) cannot always publicly endorse the CSR performance of a company, because it is too complex to enable the assessment of an individual company's performance.

THE INCORPORATION OF CSR IN NETWORKS AND CHAINS

The incorporation of CSR in networks and chains can result in intra-company and extra-company management problems if we take the development model of CSR at company level in mind (Wolfert et al. 2003). CSR only becomes relevant if a company has done its internal homework. The company often has problems implementing the CSR strategy, not only at business-unit level but also throughout the organization and down to the shop floor (Van der Schans 2004). For business development management we can look at existing business development models for CSR, such as risk management concepts and extended reporting including sustainability performance. We should not only look back but also look forward in these reports. This would provide business analysts and stakeholders with useful information for assessing a company's future risk profile.

Another interesting direction is to link up with existing quality management concepts such as the management model of the Dutch 'Instituut Nederlandse Kwaliteit' (INK model) (Wolfert et al. 2003). The advantage is that in agribusiness quality management is more widespread than external financial reporting and the INK model leaves ample room for incorporating CSR (Van Heeswijk and Schenkelaars 2002). The first steps have already been made in Van Marrewijk and Hardjono's European Corporate Sustainability Framework (Van Marrewijk and Hardjono 2003). LEI has extended this model to a chain/network approach for which also a number of tools have been developed. This will be discussed further on in this paper.

The extra-company management problems deal with the organization of cooperation (Wolfert et al. 2003). In general, chain cooperation theories largely deal only with economic performance. In chain cooperation (Goddijn and De Vlieger 2004) two aspects are of interest, namely content and process. Content is about the 'what' question: What aspects are brought into chain cooperation and what role should each party play? Process aspects deal with the 'how' questions: How do we start and develop chain cooperation, how do we work together? Important aspects of cooperation are the selection of partners with complementary competences (content), see Prahalad and Hamel (1990), and the development of trust (process), see Diederen and Jonkers (2001).

One way to establish trust is to define, standardize, control and certify each company's responsibilities and procedures. Another way is to explain and communicate each individual participant's viewpoint and strategy in relation to CSR. This increases the predictability of their behaviour. In order to start successful chain cooperation that generates added value it is not enough to look at performance only: more attention to vision and strategy is needed.

MANAGEMENT OF SOCIAL RESPONSIBILITY IN CHAINS AND NETWORKS

Managing a supply chain with respect to CSR deals mainly with overcoming problems of cooperation (competences and trust). The following two sections deal with these issues. The third section covers the development of a Chain Social

Responsibility model by combining the models of cooperation with CSR characteristics.

Cooperation and competences

A model by Doz and Hamel (1998), shown in Figure 1, integrates a number of management issues with respect to competences. This model distinguishes three phases:

1. Getting started.
2. Phasing.
3. Monitoring.

1 Analysis: 'getting started'

Finding and surmounting initial differences between partners with respect to:

- strategic aspects (goals)
- economic aspects (revenues and investment opportunities)
- operational aspects (competences and common activities)

2. Phasing

Determination of the development agenda with respect to:

- task in chain cooperation
- cooperation process
- competences
- goals and expectations

3. Monitoring

Monitoring of progress with respect to:

- results
- process.

Examples of process indicators:

Signals and signs of slow progress

Warning signal	Serious signs
Growing frustration	Scapegoats
Diminishing communication	Minimal communication
Unsolved questions	Open conflicts
Diminishing enthusiasm	Resignation

Figure 1. *Coordination of competences (source: Doz and Hamel 1998)*

Getting started is an important phase that determines the design and outcome of the coordination of competences. Three aspects are relevant:

1. The strategic basis of the cooperation. Co-operations can add value by enhancing the critical mass (get countervailing power), by developing new markets and/or products and by learning competences from each other.
2. The cooperation design. In cooperation it is important to fine-tune competences as Wolfert et al. (2003) made clear. Aspects are the contributed assets of each participant, the scope of the coordination and the common efforts.
3. Assessing and bridging differences.

To analyse and monitor competences the model of Doz and Hamel (1998) can be used. The model of Doz and Hamel is a static one. It does not tell us how cooperation develops overtime. The model of Ring & Van der Ven (1994) in Figure 2 gives more information about this aspect of cooperation. It shows a cyclical development of cooperation with three main phases: Negotiations, commitment/ obligations and realisation. In each phase the progress is constantly evaluated with respect to progress made, fair dealing and efficiency. The model distinguishes 6 important elements for the development of coordination, namely:

1. Preceding experiences (including reputation).
2. Future expectations of the cooperation.
3. Phase 1: negotiating with respect to common expectations and goals.
4. Phase 2: obligations with respect to future activities.
5. Phase 3: realization of the activities.
6. Continuous evaluation of the steps in phase 1 to 3.

In an ongoing process of cooperation the experiences during the preceding cycle of phases 1 to 3 are part of the circular process and are used to update future expectations and to answer to changed circumstances with a new round of negations, commitment and realization (phases 1 to 3).

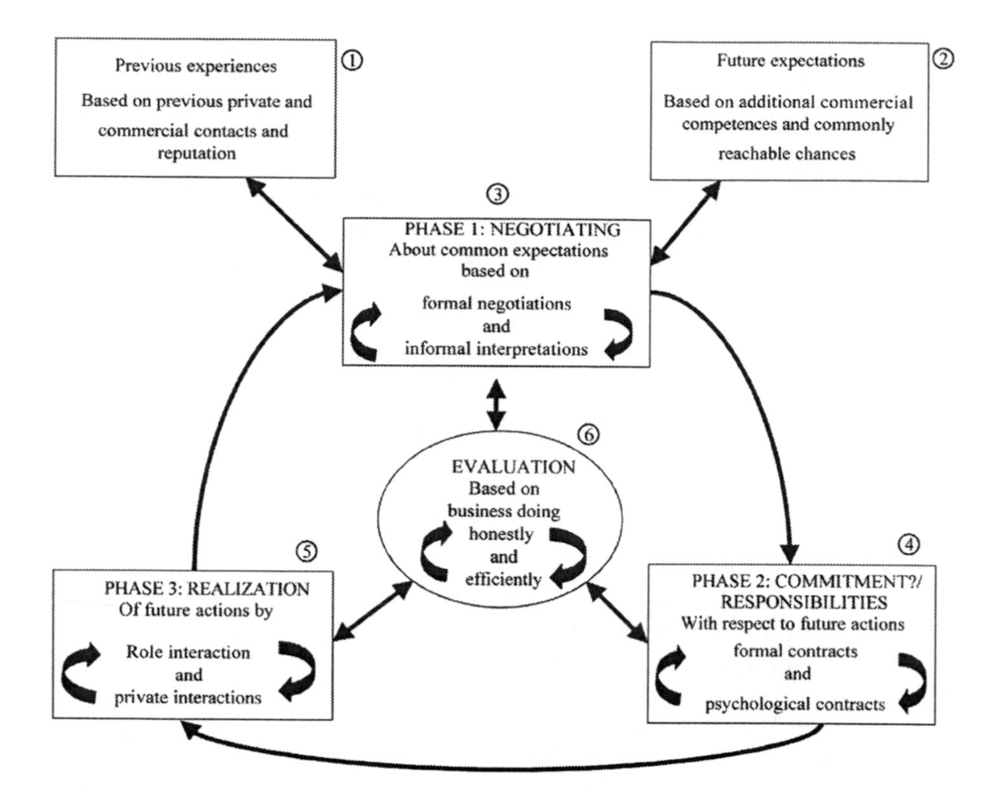

Figure 2. *Cyclical development of cooperation in supply chains. Source: Ring and Van der Ven (1994)*

Cooperation and trust

The role of trust for cooperating in chains and networks has received a lot of attention in research. In recent literature trust is considered to be a multidimensional construct present at interpersonal or system or institutional level (Lewicki and Bunker 1996). In this research the following forms of trust are distinguished:

1. Initial trust (the continuum reaches from "I don't trust anybody until it is proven they can be trusted" to "I trust everybody until it is proven they can't be trusted").
2. Trust based on cognition, which is in fact based on knowledge about the behaviour of the partner, makes the prediction of future behaviour possible.
3. Trust based on affection, which is strictly personal and mainly based on feelings.
4. Trust based on calculation, which is present if the sanction on not fulfilling the obligations of the cooperation is worse than the advantages of fulfilling them.

The model in Figure 3, based on Lewicki and Bunker (1996), gives an overview of the development of trust. The identification of trust mentioned in this model is the same as the affective trust mentioned before. The model is based on a time sequence of the different types of trust. Recent research of Klein Woolthuis (1999) showed a development of the different types of trust at the same time. It is important to mention here that trust builds up over time and that in every phase of the development of the cooperation attention should be given to the different types of trust.

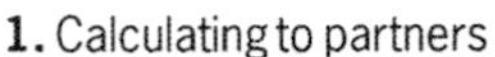

1. Calculating to partners
2. Knowledge about partners
3. Identification with partners

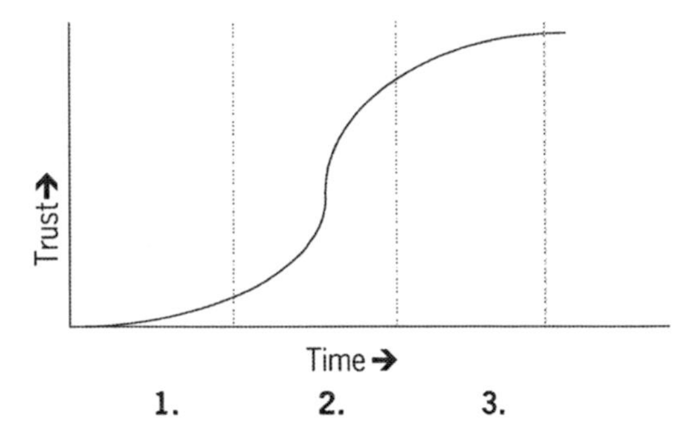

Management

Calculating	Knowledge	Identification
- sanction - possibilities - monitoring	- partners predictable - exchange of information (formal + informal)	- mutual understanding (easily sympathizing) - opencommunication

Figure 3. *The development of trust. Source: Lewicki and Bunker (1996)*

Cooperation and CSR

How can the prevailing more general models about cooperation be coupled with the more specific cooperation in the field of CSR? In a study by Goddijn et al. (2004), cooperation was integrated with the recently developed European Corporate Sustainability Framework (ESCF 2004). In the terms of Wolfert et al. (2003), a company approach was chosen and integration of CSR with quality assurance models was applied.
The ECSF framework is meant to help organizations to structure their CSR approach with respect to adapting, connecting and starting CSR in the organization itself. CSR has a multidisciplinary nature and needs a holistic approach. The SqEME model (Figure 4; ESCF 2004) offers a transparent design to deal with the complexity of CSR. SqEME focuses on information about: 1. the direction (constitution); 2. the norms for intervention (chemistry); 3. the actions (conduct); and 4. performances (control).

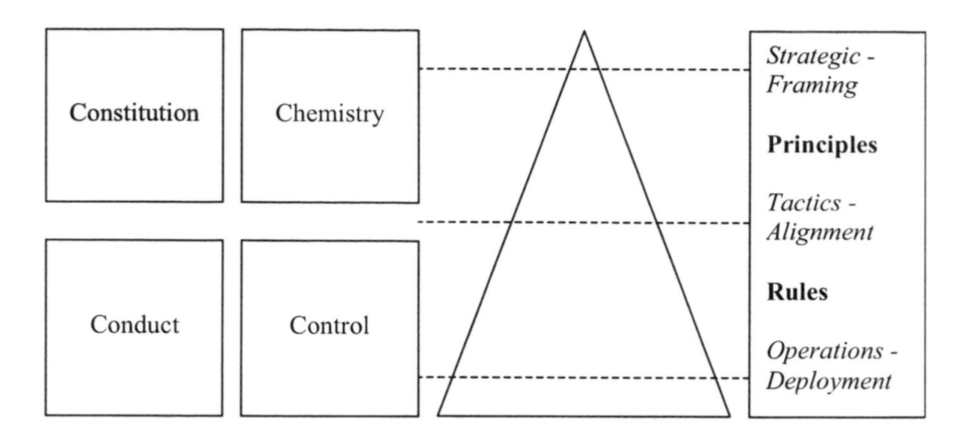

Figure 4. *The SqEME model. Source: ECSF (2004)*

In the SqEME analysis, the objectives of the organization are determined first, by looking at values, identity, mission and vision (constitution). Secondly the factors that affect the success of the chosen direction (chemistry) are addressed. The elements analysed are energy, streams, relations and communications. These are the factors that are considered to make the organization work. The focus 'conduct' is oriented to what is really happening in the organization (organizational behaviour, processes and procedures). Control is about reflection and learning, which can be measured and monitored. It gives us an evaluation of the CSR policy and practise of the firm.

Another part of the ECSF (2004) framework distinguishes a strategic (framing), tactical (alignment) and operational (deployment) level. In the SqEME model a distinction between principles and rules is made. Alignment means that corporate strategies, policies and leadership on CSR can manifest themselves in people management, resource management and organizational processes. The three levels in the ECSF model are the same as distinguished in the INK model (Van Heeswijk and Schenkelaars 2002): aiming, arranging and performing. It enables the introduction and coordination of different time horizons.

Furthermore, the ECSF framework is supported by theory and by instruments on corporate sustainability. For the 'constitution' focus the Emergent Cyclical Levels of Existence Theory (ECLET) of Graves is used (ESCF 2004). The 'chemistry' focus is supported by the Four Phase model of Hardjono et al. (2004), and the focuses 'conduct' and 'control' by the EFQM excellence model/INK model (2002) and the Business Balanced Scorecard of Kaplan and Norton (2001).

The ECLET model provides a framework to handle subjects that are difficult to define, like CSR. Each interpretation results in a different understanding of external development and organizational solutions. The Four Phase model is a strategic tool to understand organizational rhythm and dynamics (Hardjono and Rouppe van der Voort 2004). The strategic orientations are based on the internal or external orientation and on a control or change orientation. The following orientations are distinguished: effectiveness/market, efficiency/production, flexibility and creativity/innovativeness.

The EFQM model (www.efqm.org/model) supports organizations at the level of alignment and tactics. The enabling instruments such as leadership, strategy and policy, people, resources and processes are used in the 'conduct' focus and the four result areas for the 'control' focus. On the operational level the Balanced Scorecard is used, which assesses the following well-known four perspectives: learning and growth, business process, clients, and financial performance.

With the help of the self-assessment tool of ECSF, organizations can get information about the integration of CSR in the organization as a whole. This assessment starts with the strategic principles of CSR (constitution and chemistry focus), then the assessment of CSR rules is made (conduct and control focus) and, finally, comparing the profiles assesses the results.

The ECSF framework and the connected assessment tool for CSR give a good overview of the present situation. However, they can also be used as a tool to determine the desirable situation.

CSR in supply chains

Based on the theories, tools and models discussed previously in this paper, LEI developed a conceptual model for CSR in chains (Figure 5). This model integrates the models for chain cooperation and the ECSF framework. The model of Ring and Van de Ven (1994) offers instruments to handle the dynamics in cooperation. The model of Doz and Hamel (1998) gives the general content of the cooperation, while the ECSF framework (ESCF 2004) gives the specific content of CSR.

The implementation of the developed model starts with a number of work sessions with possible chain partners to develop a common strategy that gives the cooperation a goal/direction. This strategy will be based on mission, competences, external and internal situation, strengths and weaknesses of the partners. The sessions not only result in information about the direction of the cooperation (constitution) but also in information of importance for the manner of cooperation itself (chemistry). With respect to competences the scope of the cooperation is of importance and also the added value of each of the partners in the alliance.

During the first session the chain partners have to sketch their present situation with respect to constitution, chemistry, competencies and added value. Trust is of importance in every phase of the cooperation. In the work sessions this is addressed with the help of firm presentations, clear arrangements and common discussion about a monitoring system with respect to the cooperation.

After the partners have made an agreement about mission, strategy and competences, the second phase starts. This phase is meant to create commitment about obligations and actions within the cooperation with respect to guidelines, performance criteria and reporting. This results in a cooperation agenda. Then the actions agreed upon are followed up (conduct focus). The evaluation of the cooperation is a continuous task (control focus).

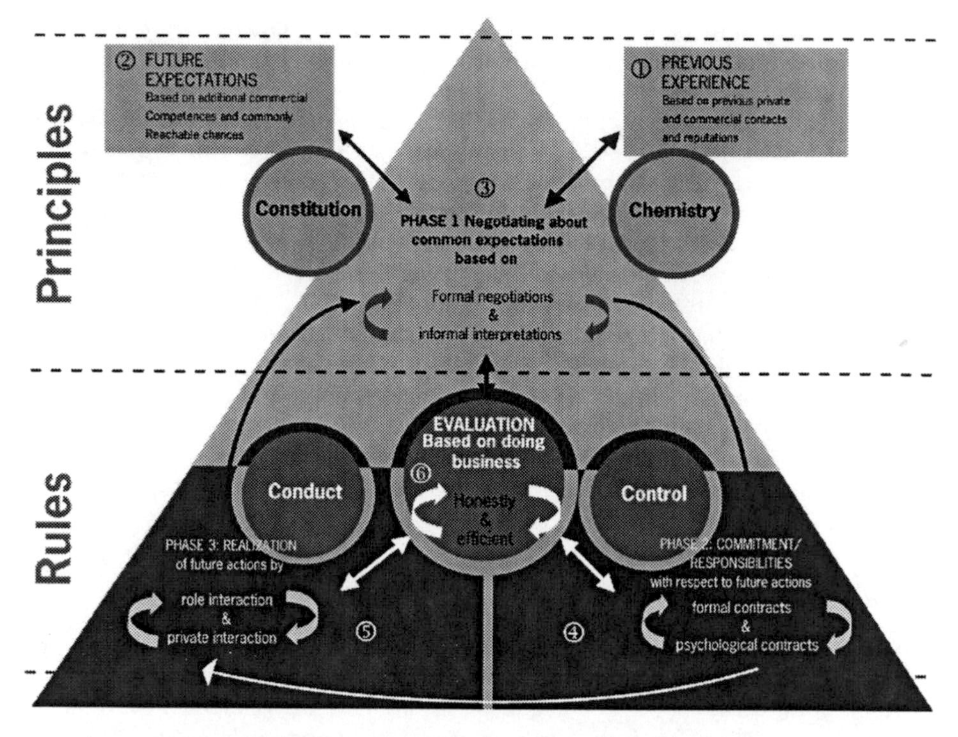

Figure 5. *An integrated model to analyse and manage CSR in chains and networks. Source: Ring and Van der Ven (1994) and ECSF (2004) adapted by LEI*

CONCLUSIONS

CSR involves both a content (what) and a process (how) aspect in which principles, processes and performances play a part. It can be seen as a bundle of 'credence' properties of a product. To avoid information asymmetry in the market, transparency of the production process is needed. For agriculture and agribusiness a company-oriented approach to transparency is the most suited, given the importance of by-products and externalities in the production process. It gives companies the opportunity to prove they are producing in a socially responsible way.

Network and Chain Social Responsibility is the final stage in a development process of CSR within a firm from the stage 'put in order' to 'integrate'. Given this sequence of stages partners have to prepare themselves for this approach before Chain Social Responsibility is possible. Using a combination of models is important for the analysis and understanding of chain cooperation and Chain Social Responsibility. These models are the 'coordination of competences' model by Doz and Hamel (1998), the 'cyclical development of cooperation in supply chains' model by Ring and Van de Ven (1994), and the 'development of trust' model by Lewicki and Bunker (1996). The combination of these models with the ECSF approach is a promising tool to facilitate the development of CSR in a chain or network and manage it in a sustainable manner.

NOTES

[1] For a 'credence' product it is not possible to determine the quality of it before or after the purchase. For a 'search' good the quality can be determined before the purchase and for an 'experience' good this can be done after the purchase (Kola and T. 2003).

REFERENCES

Akerlof, G.A., 1970. The market for lemons: quality uncertainty and the market mechanism. *Quarterly Journal of Economics,* 84 (3), 488-500.

Antle, J.M., 1999. The new economics of agriculture. *American Journal of Agricultural Economics,* 81 (5), 993-1010.

Becker, T., 2000. A framework for analyzing public and private food quality policy: meeting consumer requirements. *In:* Becker, T. ed. *Quality policy and consumer behaviour in the European Union.* Wiss.-Verl. Vauk, Kiel, 91-110.

Berenschot, 2003. *Maatschappelijk verantwoord ondernemen in de etalage.* Berenschot Groep B.V., Utrecht.

Brundtland, G. (ed.) 1987. *Our common future: the World Commission on Environment and Development.* Oxford University press, Oxford.

Carroll, A.B., 1979. A three-dimensional conceptual model of corporate performance. *Academy of Management Review,* 4 (4), 497-505.

Caswell, J.A. and E.M., Modjuzska, 1996. Using informational labelling to influence the market for quality in food products. *American Journal of Agricultural Economics,* 78, 1248-1253.

Commission of the European Communities, 2002. *Corporate social responsibility: a business contribution to sustainable development.* Commission of the European Communities, Brussels. Com(2002) 347 final.

De Ron, A., Keijzers, G., Boons, F.J., et al. 2002. *Duurzaam ondernemen. Strategie van bedrijven.* Kluwer, [Alphen aan den Rijn].

Diederen, P.J.M. and Jonkers, J.L., 2001. *Chain and network studies.* KLICT, 's-Hertogenbosch. KLICT/2415/RdG. [http://www.klict.org/docs/pp2415.pdf]

Doz, Y.L. and Hamel, G., 1998. *Alliance advantage: the art of creating value through partnering.* Harvard Business School Press, Boston.

ESCF, 2004. *Self assessment on corporate sustainability, draft version.* Erasmus Universiteit, Rotterdam.

Goddijn, S.T. and De Vlieger, J.J., 2004. *MVO in ketens: een MVO-ketensamenwerkingsmodel en een voorstel tot ontwikkeling van een GRI-jaarrapport voor de foodsector.* Lei, Den Haag. Projectcode 30054 [http://www.lei.wageningen-ur.nl/publicaties/PDF/2004/5_xxx/5_04_09.pdf]

GRI, 2002. *Richtlijnen voor duurzaamheidsverslaggeving.* Global Reporting Initiative, Amsterdam.

Hardjono, T.W. and Rouppe van der Voort, M.B.V., 2004. *The four-phase model, maturity of organizations and business excellence.* Erasmus Universiteit, Rotterdam.

Kaplan, R.S. and Norton, D.P., 2001. *Focus op strategie.* Business Contact, Amsterdam.

Klein Woolthuis, R.J.A., 1999. *Sleeping with the enemy: trust, dependence and contracts in interorganisational relationships.* Proefschrift Universiteit Twente, Enschede

Kola, J. and T., Latvala, 2003. *Impact of information on the demand for credence characteristics: selected paper in the IAMA International Food and Agribusiness Management Association world food and agribusiness symposium and forum, June 21-24, 2003, Cancun, Mexico.* IAMA, College Station. [http://www.ifama.org/conferences/2003Conference/papers/kolaIMPACT.pdf]

Lewicki, R.J. and Bunker, B.B., 1996. Developing and maintaining trust in work relationships. *In:* Kramer, R.M. and Tyler, T.R. eds. *Trust in organizations: frontiers of theory and research.* Sage, Thousank Oaks, 114-139.

Meeusen, M.J.G. and Ten Pierick, E., 2002. *Meten van duurzaamheid: naar een instrument voor agroketens.* Lei, Den Haag. [http://www.lei.nl/publicaties/PDF/2002/5_xxx/5_02_11.pdf]

OECD, 2001. *The OECD guidelines for multinational enterprises: text, commentary and clarifications*. OECD, Paris. [http://www.olis.oecd.org/olis/2000doc.nsf/4f7adc214b91a685c12569fa005d0ee7/d1bada1e70ca5d90c1256af6005ddad5/$FILE/JT00115758.PDF]

Pannebakker, R. and Boone, J.A., 2004. *Duurzaamheidsverslaggeving in de Nederlandse agrosector: een empirisch onderzoek over 2001 en 2002*. Lei, Den Haag. LEI Rapport no. 5.04.05. [http://www.lei.wageningen-ur.nl/publicaties/PDF/2004/5_xxx/5_04_05.pdf]

Prahalad, B. and Hamel, G., 1990. The core competence of the corporation. *Harvard Business Review*, 63 (3), 79-91.

Ring, P. and Van de Ven, A.H., 1994. Developmental processes of cooperative interorganizational relationships. *Academy of Management Review*, 19 (1), 90-118.

SER, 2000. *De winst van waarden: advies over maatschappelijk ondernemen*. SER, Den Haag. [http://www.ser.nl/_upload/databank_adviezen/B19054.pdf]

Udo de Haes, H. and Sonnemann, G., 2003. *Life Cycle Initiative: task forces: overview and terms of reference. Final draft*. UNEP/SETAC. [http://www.uneptie.org/pc/sustain/reports/lcini/ToR_overview_Draft%20Final_.pdf]

Van der Schans, J.W., 2004. *Varkensketens in transitie*. Lei, Den Haag. [http://www.lei.dlo.nl/publicaties/PDF/2004/1_xxx/1_04_03.pdf]

Van der Schans, J.W., Vogelzang, T.A. and De Vlieger, J.J., 2003. *Maatschappelijk verantwoord ondernemen in de agrofood keten: in het bijzonder in de zuivelsector*. Lei, Den Haag. LEI Rapport no. 2.02.15. [http://library.wur.nl/wasp/bestanden/LUWPUBRD_00320864_A502_001.pdf]

Van Heeswijk, H.A.G.M. and Schenkelaars, E., 2002. *Handleiding positiebepaling ondernemingen*. INK, Zaltbommel.

Van Marrewijk, M. and Hardjono, T.W., 2003. European corporate sustainability framework for managing complexity and corporate transformation. *Journal of Business Ethics*, 44 (2/3, pt. 1/2), 121-132.

Wolfert, J., Kramer, K.J. and Van der Schans, J.W., 2003. *Implementing corporate social responsibility into new chains in a transparent way: opportunities and challenges*. KLICT, 's-Hertogenbosch. KLICT Position Paper no. TR171/3.

Wood, D.J., 1991. Corporate social performance revisited. *Academy of Management Review*, 16 (4), 691-718.

Wood Renck, A., 2002. *Credence goods: a labelling problem? Paper presented in annual meeting Southern Agricultural Economics Association, Mobile, Alabama, Febr. 1-5, 2003*.

CHAPTER 15

GOVERNANCE STRUCTURES IN THE DUTCH FRESH-PRODUCE INDUSTRY

JOS BIJMAN

Department of Business Administration, Wageningen University, Hollandseweg 1, 6706 KN Wageningen, The Netherlands. E-mail: jos.bijman@wur.nl

Abstract. A governance structure is the set of public and private rules that govern the execution of a transaction. Governance structures affect the efficiency of transactions by solving two basic problems of exchange: coordination and safeguarding. Coordination refers to the alignment of the activities of two or more parties involved in the same transaction. Safeguarding refers to protecting against exchange hazards such as shirking and hold-up. This paper presents a model for studying governance structure choice. The model goes beyond traditional conceptualizations of governance structure by identifying the governance mechanisms that solve the safeguarding and coordination problems. The model is applied to changes in governance structures in the Dutch fresh-produce industry.
Keywords: governance structure; transaction costs; safeguarding; coordination; fresh produce

INTRODUCTION

For a long time, one governance structure dominated the marketing of fruits and vegetables in The Netherlands: the cooperative auction. Recently, this governance structure has lost much of its appeal, while new governance structures have become more popular. This paper tries to explain both the long-time popularity of the cooperative auction and the recent growth in different sales structures. It uses concepts from economic organization theory, organization theory and social-network theory. The paper develops an integrated model for studying governance structure change and presents a first application of the model to the changes in the Dutch fresh-produce industry over the last decade. The model goes beyond traditional conceptualizations, which treat a governance structure as a black box. Our model specifically targets the two main functions of every governance structure – safeguarding and coordination. It focuses on the various governance mechanisms that can be used to solve the problems of safeguarding and coordination. In doing so, it also acknowledges the distinction between formal and informal mechanisms.

The paper starts with a brief explanation of the causes of transaction costs, and with answering the question: what is a governance structure? Then, the two main

C.J.M. Ondersteijn et al. (eds.), Quantifying the agri-food supply chain, 207-223.

functions of each governance structure – safeguarding and coordination – are discussed. This review of the literature results in the theoretical model. After a brief description of the marketing channels in the Dutch fresh-produce industry, the model is applied to explain the popularity of the dominant governance structure, the grower-owned cooperative auction. Next, the shifts in governance structure in the last decade will be described and explained with the model. We will finish with conclusions on the applicability of the model.

TRANSACTION COSTS AND GOVERNANCE STRUCTURE CHOICE

The concept of governance structure comes from institutional economics, as developed by Coase (1937), Klein et al. (1978), Williamson (1979; 1987; 1991), Barzel (1982), Cheung (1983) and many others. Central in institutional economics is the notion that costless exchange between any two of more economic agents (persons, firms or organizations) does not exist. Any transaction will come with costs for the agents: transaction costs.

Transaction costs are the costs of contact, contract and control, i.e., the costs associated with finding a market and a trading partner, negotiating an agreement, and monitoring and enforcing the contract. Transaction costs are caused by the particular characteristics of a transaction. In the economic organization literature we find at least five characteristics of transactions that affect the size of transaction costs: asset specificity, uncertainty, frequency, measurement problems, and connectedness to other transactions.

Williamson (1979; 1987) distinguishes three characteristics: the presence of transaction-specific assets (i.e., asset specificity), the uncertainty surrounding or the complexity of the transaction, and the frequency of the transaction. Other authors have added the difficulty of performance measurement, and the connectedness of a transaction to other transactions. The difficulty of performance measurement (because of information asymmetry) is a typical agency problem (Barzel 1982; Holmström and Milgrom 1991; 1994). A fifth potential cause of transaction costs has been added by Milgrom and Roberts (1992): the connectedness of a transaction to other transactions carried out by other parties. The notion of connectedness in organizational economics is the same as the notion of interdependence in traditional organization theory.

Transaction Cost Economics (TCE) posits that when transaction cost are low, the transaction will be carried out through the governance structure *spot market*, and when transaction costs are high, it becomes efficient to set up an organizational structure (*hierarchy* in the terminology of Williamson) for carrying out the transaction. In between market and hierarchy, there is the governance structure *hybrid*. Williamson (1991) emphasizes the discreteness of governance structures. Cheung (1983), building on Coase (1937), has developed the notion of a continuum of governance structures.

When transaction costs increase or decrease, a different governance structure may be chosen to carry out the transaction. Such a shift in governance structure, either from one discrete form to another or along a continuum, raises the question what is actually changing. Which attributes of a governance structure may change when one or more of the five characteristics of transactions change? Answering this question requires a closer look at the constituent elements and the functions of a governance structure.

WHAT IS A GOVERNANCE STRUCTURE?

Williamson (1979) defined a governance structure as "the institutional framework within which the integrity of a transaction is decided". Other authors, applying or further developing TCE, have used governance mechanism or governance form instead of governance structure. For instance, Hesterley et al. (1990, p. 403) provide the following definition: "a governance mechanism includes any institutional arrangement that serves to influence the exchange process". Besides by the private institutional arrangement, transactions are also governed by the institutional environment, such as laws and rules that apply beyond the specific transaction. These laws and rules may also be considered parts of the governance structure.

In institutional economics, the emphasis is on formal institutions, such as laws, contract rules, formal codes of conduct, and official arrangements, which together make up the governance structure. However, informal institutions, such as norms, traditions, customs and culture, also influence transactions. The role of informal institutions in supporting transactions has been emphasized by social theorists, studying the embeddedness of exchange relationships. This embeddedness has two dimensions: relational and structural (Granovetter 1992). Relational embeddedness refers to the ongoing social relationship that results from repeated transactions with the same partner. Structural embeddedness refers to the fact that the dyadic relationship is embedded in a community of former, current and potential exchange partners. Being part of a community, where information on individual behaviour is exchanged, leads to a reputation effect. A central theme in the research on embeddedness is that repetitive market relations and the linking of social and business relationships generate embedded logics of exchange that differ from those emerging in traditional arms-length market relations (Borgatti and Foster 2003). Social mechanisms, such as reputation, restricted access, macroculture and social sanctions, are important elements of network governance (Jones et al. 1997).

We conclude that two perspectives on governance structures exist. Institutional economists focus on the formal institutions, while social-network theorists use a broader definition by also including informal institutions. In this paper we adhere to the broad approach, and define a governance structure as the set of formal and informal institutions that regulate a particular transaction.

SAFEGUARDING AND COORDINATION

To understand changes in governance structure it is not sufficient to know what a governance structure is; it is also necessary to know what a governance structure does. TCE posits that a governance structure is chosen in order to economize on transaction costs. But how does a governance structure support efficient transactions? Williamson (1999) emphasizes three basic elements of all transactions: mutuality, conflict and order. "Governance is a means by which to infuse *order* in a relation where potential *conflict* threatens to undo or upset opportunities to realize *mutual* gains" (Williamson 1999, p. 1090). This means that a governance structure furthers the efficiency of a transaction by supporting the realization of mutual gains and by preventing or solving conflicts. Putting it differently, we may state that the two main functions of a governance structure are coordination (to obtain the mutual gains) and safeguarding (to avoid conflict and premature termination of the agreement).

The problems of coordination and safeguarding are usually studied from different theoretical perspectives. Theories that emphasize the need to safeguard an agreement start from the assumption that parties have conflicting interests and therefore may behave opportunistically, taking advantage of a situation of asymmetric information. Theories emphasizing the need for coordination assume corresponding interests, but acknowledge the bounded rationality of human actors. These theories focus on the methods to solve the problems of incomplete or asymmetric information between transaction partners.

Most economic approaches to efficient governance structure choice have focused on the safeguarding function. Typically, in TCE it is claimed that behavioural uncertainty and bounded rationality lead to contractual hazards such as the appropriation of quasi-rents by one of the transaction parties in situations with asset specificity. Also measurement problems in transactions lead managers to choose a governance structure that minimizes the transaction costs caused by the combination of incomplete/asymmetric information and incomplete commitment. In TCE, it is control over particular assets that provide protection against the threat of hold-up. Thus, a governance structure helps to safeguard investments because it contains a particular distribution of property rights and therefore provides formal control over the deployment of assets. Another economic approach that studies the safeguarding problem is the new property-rights theory, developed by Grossman and Hart (1986) and Hart and Moore (1990). This theory of the firm starts from the assumption that all contracts are incomplete and that lock-in or hold-up may develop when investments are relationship-specific. Ownership of particular assets may prevent ex-post lock-in or hold-up, because the owner of an asset is in a good position to bargain over the deployment of that asset. Expecting the risk of ex-post appropriation, contract partners may take suboptimal ex-ante investment decisions. By shifting the ownership of specific assets, the efficiency of the transaction can be improved.

So far, we discussed the safeguarding problem. However, firms choose governance structures not only to address appropriation concerns, but also to manage anticipated coordination problems. Coordination costs are the costs of information processing and decision-making that result from decomposing tasks between partners to an exchange. Coordination costs and the mechanisms for dealing with these costs have traditionally been object of study in organization theory. Focusing on the organization of activities within firms, organization theory emphasized the role of hierarchical controls in reducing coordination costs.

Interorganization coordination costs arise in transaction relationships where partners have agreed upon a division of labour and have to coordinate and manage, across organizational boundaries, activities to be completed jointly or individually. Within organization theory, a fundamental principle defining the costs of coordination within organizations is the concept of interdependence (Thompson 1967). An organization is dependent on some element of its task environment (1) in proportion to the organization's need for resources or performances that that element can provide and (2) in inverse proportion to the ability of other elements to provide the same resource or performance (Thompson 1967, p. 30). When dependence is mutual, it has been named interdependence. Interdependence can be symmetric, but is more likely to be asymmetric.

Thompson (1967) distinguishes three types of interdependency: pooled, sequential and reciprocal. With pooled interdependence, each part of an organization renders a discrete contribution to the whole and each is supported by the whole. The parts are interdependent in the sense that unless they perform adequately, the total organization is jeopardized. With sequential interdependence, the output of one part is the input for another part. Parts that experience sequential interdependence are also interdependent in a pooled way. With reciprocal interdependence the output of each part becomes input for the other parts. In other words, each part poses contingency for the others. Reciprocal interdependence also includes pooled and sequential interdependence.

There are three types of coordination mechanisms associated with the three types of interdependency (Thompson 1967): with pooled interdependence, coordination by standardization is appropriate; with sequential interdependence, coordination by plan is appropriate; and with reciprocal interdependence, coordination by mutual adjustment is called for. The three types of coordination, in the order presented here, place increasingly heavy burdens on communication and decision. Mintzberg (1979) has elaborated on these coordination mechanisms within organizations. Although Thompson wrote about intra-firm interdependencies, his typology can also be applied to inter-firm interdependence situations. Of course, in inter-organizational transactions, also prices function as coordination mechanisms.

The theoretical framework relating particular coordination mechanisms with particular governance structures is still in an embryonic stage. Few studies exist that explicitly develop propositions about this relationship. Exceptions are Grandori (1997) and Gulati and Singh (1998). Both use Thompson's typology of interdependencies as a proxy for coordination costs. For instance, Gulati and Singh (1998), in their study on the relationship between coordination costs and the choice of governance structure in inter-firm alliances, propose that when the nature of interdependence shifts from pooled to sequential to reciprocal, coordination costs increase. The authors have operationalized interdependence on the basis of the value-creating logic(s) of each alliance. Particular value-creating logics result in particular interdependencies and thereby lead to specific coordination costs.

In sum, each governance structure has two main functions: safeguarding the transaction from appropriation of the quasi-rents, and coordinating the activities and decisions among the transaction partners. In situations of conflicting interests, a governance structure is meant to avoid conflicts or provide solutions to a conflict.

Even when interests perfectly correspond, transaction costs arise because time and effort must be spent on information exchange and decision-making.

A THEORETICAL FRAMEWORK

So far, we have discussed the economic and organization literature about governance structure, its functions and its attributes. Five sets of characteristics determine transactions costs and thereby the choice of governance structure. Governance structures have two main functions: safeguarding and coordination. These functions are obtained through various formal and informal mechanisms (or institutions). In this section we will try to bring these elements together in a comprehensive model for studying shifts in governance structures (Figure 1).

The five characteristics of transactions determine the problems of safeguarding and coordination. To solve these problems, specific governance mechanisms are used. The problem of safeguarding investments is mainly determined by asset specificity and measurement problems. The problem of coordination is mainly determined by the frequency of the transaction, the uncertainty and complexity of the transaction, and the interdependence of the transaction with other transactions.

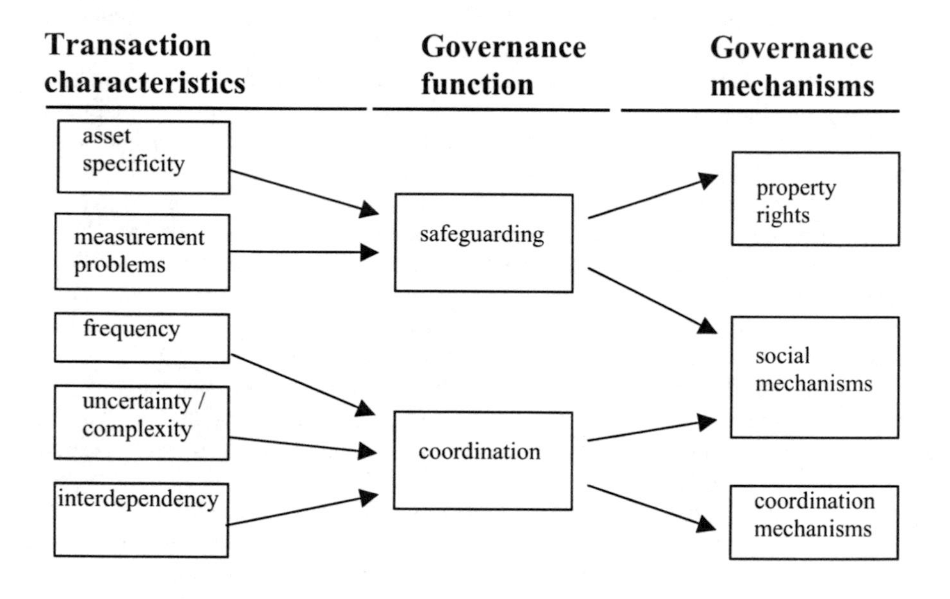

Figure 1. *Governance functions and governance mechanisms*

Each governance structure consists of a specific set of governance mechanisms, such as a particular distribution of property rights, particular social mechanisms and coordination mechanisms. Social mechanisms are informal mechanisms, while property rights and coordination mechanisms are formal mechanisms. For safeguarding, the most important formal mechanism is administrative control based on property rights. It is the owner of an asset who can decide about deployment of and access to that asset, both in the ex-ante and ex-post situation. When asset specificity increases or when performance measurement becomes more difficult, a different distribution of property rights may provide the necessary safeguarding. Informal mechanisms such as reputation and social control may also provide protection against opportunistic behaviour. For coordination, several formal mechanisms are available, such as standardization, direct supervision and mutual adjustment. For frequent transactions, standardization is the most appropriate mechanism. For transactions with high uncertainty and/or complexity, the appropriate coordination mechanism may be direct supervision or mutual adjustment, depending on the distribution of property rights. For transactions that are interdependent with other transactions, several coordination mechanisms can be used, depending on the type of interdependence. If the interdependence is of a pooled kind, standardization may be sufficient to obtain coordination. If the interdependence is of sequential kind, direct supervision may be more appropriate (besides standardization). If interdependence is of reciprocal kind, mutual adjustment may be the appropriate mechanism. Also informal mechanisms such as restricted access and cultural homogeneity may provide the information exchange and support for decision-making that are needed for coordination.

Informal institutions are not easily established and take a long time to materialize. Changes in informal institutions will become effective only after some time. For this reason, in this paper we do not take into account the working of informal institutions, but focus on formal governance mechanisms. Our model leads to the following propositions about the relationship between characteristics of the transaction and mechanisms of governance used for that transaction:

- Proposition 1: increasing asset specificity and/or measurement problems (which imply higher transaction costs) will lead to a change in the distribution of property rights.
- Proposition 2: increasing frequency, uncertainty and interdependence (which imply higher transaction costs) will lead to a shift in coordination mechanisms.

We will now use this theoretical model to assess the governance structures used in the marketing channels for Dutch fresh produce. In doing so, we will focus on the transaction between the grower and his customer (usually a trader).

MARKETING CHANNELS FOR FRESH PRODUCE IN THE NETHERLANDS

In marketing channels for fresh produce, we can distinguish at least four parties on the basis of the main functions in the channel: grower, cooperative, wholesaler and retailer. The individual grower specializes in producing fruits or vegetables; the grower-owned cooperative takes care of collection and marketing of the growers' products; the wholesaler (or general: trader) is the party that takes care of shipping the products to domestic and foreign customers, and has both a trading and logistic function; and the retailer, with his distribution function, is the gateway to the consumer.

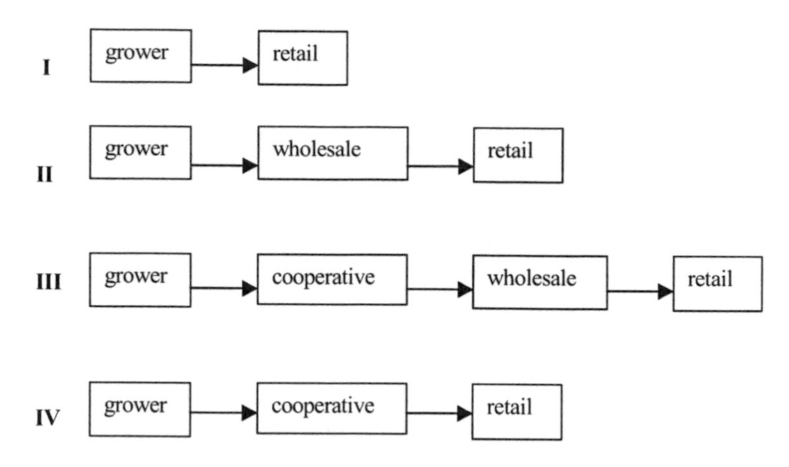

Figure 2. *Four marketing channels for fresh produce*

For the marketing of their products, growers can choose between different channels, involving two or more of the four main parties (Figure 2). The first marketing channel (I) consists of only two parties: an individual grower sells directly to a retailer. This channel does exist in reality, but is not very common because of scale differences between production and retail, because most retailers prefer to deal with only a limited number of suppliers and not with each producer individually, or because the grower does not want to perform the sales function himself. The second channel (II) consists of grower, wholesaler and retailer. The wholesaler has a collection function as well as a distribution function, buying from growers and selling to retailers. In the third channel (III), growers have delegated the collection and marketing function to a grower-owned cooperative. As the marketing function is carried out by the cooperative, growers collectively benefit from economies of scale and scope in marketing as well as from stronger bargaining power, while individually they can specialize in production. In the fourth channel (IV), the wholesale function is also carried out by the grower-owned cooperative. In this channel, there is no independent wholesaler, as the cooperative directly trades with retailers.

In practice there may be more marketing channels, longer channels or more complex channels. In the case of export, the channel is longer, because there usually is an exporting trader on one side of the border and an importing trader on the other side. For our exposition it is sufficient to distinguish these four main parties.

THE TRADITIONAL DUTCH MARKETING CHANNEL FOR FRESH PRODUCE

Traditionally, the dominant marketing channel for fruits and vegetables in The Netherlands is model III. Growers bring their products to the cooperative auction, where they are sold to wholesalers and to some retailers. Wholesalers sell to domestic retailers or export the products to foreign wholesalers or retailers. Retail takes care of selling to the final consumer. The main function of the grower-owned cooperative is to provide an organized marketplace for the growers to sell their products. This service included running the auction clock for price determination, sales administration on behalf of the sellers, logistic services (mainly short-term warehousing), and quality classification and inspection (as uniformity in products supports the sales process) (Meulenberg 1989). For almost a century, the cooperative auction was the most popular marketing channel for fresh produce in The Netherlands.

In the days before the auction, in the 19th century, different marketing channels were used (Kemmers 1987). Growers located in the vicinity of cities sold their products directly to retailers and local traders. Growers in remote areas sold their products to traders, who shipped the produce to the main cities of Holland or neighbouring countries. Several growers had established export associations, hiring sales personnel to find customers abroad. These sales methods had several disadvantages, such as information asymmetry between grower and independent trader, agency problems in monitoring the effort of sales personnel, and high logistic costs because of multiple stages in the marketing channel.

The auction method for selling vegetables was used for the very first time in 1887; while in 1889 the first organized auction was established (Kemmers 1987). The real breakthrough in the popularity of the organized auction for selling fruits and vegetables occurred in the first decade of the 20th century (Kemmers 1987). Because of economic prosperity in Northwest Europe, demand for fresh produce was growing. Growers felt that the traditional marketing structures were insufficiently equipped to exploit the growing demand (Van Stuijvenberg 1977). The auction became popular because of the speed of the sales process, the opportunities for new traders to compete with incumbent firms, and the transparency of the market.

In the early decades of the 20th century, a large number of auctions were established. Every region with professional horticulture set up its own cooperative auction. While exact numbers for the years in between 1890 and 1915 do not exist, Kemmers (1987) gives anecdotal evidence of the rapid increase in the number of auctions in the early years of the 20th century. Figure 3 shows that within 25 years more than 120 new auctions were set up. In 1934 an Auction Law was enacted as part of government policy to alleviate the effects of the economic crisis of the 1930s. This law contained a legal obligation for growers of fresh produce to sell their products through an auction. In 1945 the total number of fresh-produce auctions reached its top with 162 (Plantenberg 1987). After World War II, the number of auctions gradually declined, mainly due to mergers of cooperatives, in order to gain economies of scale. The fastest decrease in the total number of auctions occurred after 1965, when the auction law was abolished. While other marketing channels were becoming more popular, the cooperative auction remained the most dominant one. Between 1965 and 1995 the share of all fruits and vegetables being sold through an auction declined from 100 to 75 percent (Bijman 2000). In 2000, only six cooperative auctions remained.

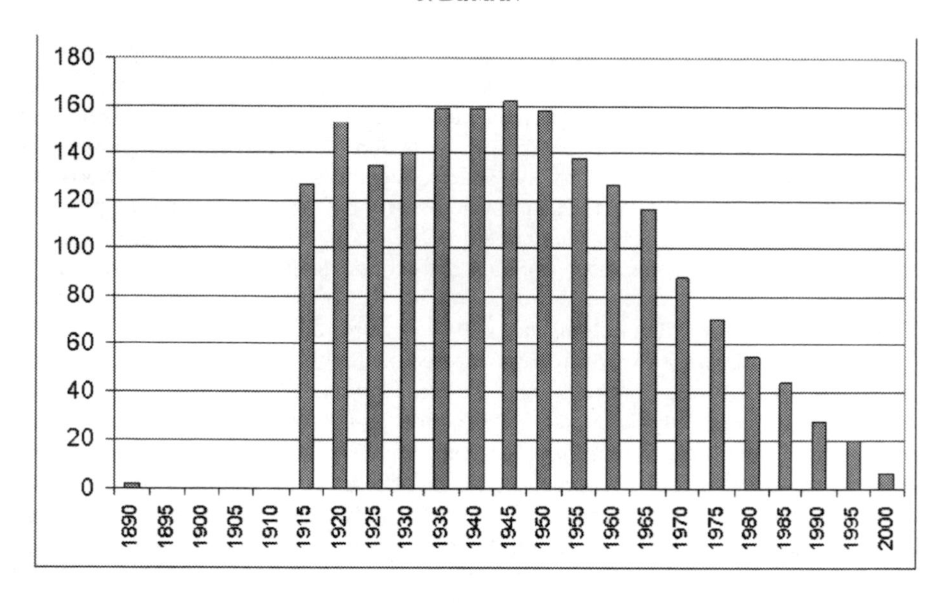

Figure 3. *Number of vegetable auctions in The Netherlands, 1890-2000*

THE COOPERATIVE AUCTION AS DOMINANT GOVERNANCE STRUCTURE

According to TCE logic, the popularity of a particular governance structure is an indication of its efficiency, as over time efficient structures out-compete less efficient structures. Also organizational ecology (Hannan and Freeman 1989) argues that efficient organizational forms will survive in an evolutionary process of variation, selection and retention. With the theoretical framework developed above we can explain why growers chose the cooperative auction as the favourite governance structure.

Traditional transactions between growers of fruits and vegetables and their wholesale customers had the following characteristics:

- moderate asset specificity: temporal asset specificity because of the perishability of the products; moderate site specificity when transportation costs are high; no physical assets that can result in bilateral dependency
- high frequency, because harvested products must be sold immediately
- low uncertainty, as long as products are generic
- high measurement problems in case growers individually contract with a trader
- no interdependence, unless growers collectively sell their products, in which case pooled interdependence is present.

Pure market governance is not an efficient option for growers, because of the transaction costs that result from temporal asset specificity and asymmetric information between grower and trader. Transaction costs are also quite high because growers have to spend time and effort on investigating and following demand conditions, which are volatile because of the variation in both supply and demand. Transaction costs also result from traders opportunistically using their information advantage. The other extreme, a hierarchical governance structure, is not an efficient option either, as individual growers do not have the scale and knowledge to carry out the wholesale function and traders do not have the idiosyncratic knowledge of the production conditions. Substantial problems of performance measurement would arise when a grower hires a salesperson or when a wholesaler hires a grower. In fact, growers seek a governance structure that will let them specialize in production, outsource the marketing function and benefit from the scale economies in marketing, while avoiding the agency problems that usually come with contracting an independent marketing firm.

The cooperative auction was the efficient governance structure growers were looking for. Information costs are low because of the high transparency of the market. Buyers come to the auction, and the auction clock determines prices. Contract costs are non-existent, because no contract negotiation is needed between sellers and buyers. Monitoring and control costs are very low, because the transaction between grower and buyer is close to a pure market transaction. Thus, no agency problems (i.e., performance measurement problems) are present in the transaction. As growers have to be members of the cooperative and buyers have to be registered, compliance to the auction rules is guaranteed by private order. Moreover, the cooperative provides collective insurance to the growers against buyer default.

The high frequency of transaction and the economies of scale justify the establishment of a particular organization. But why was the auction set up by the growers, and not by buyers or by an independent firm? The explanation comes from the specificity of the auction assets. Site specificity, physical asset specificity and temporal asset specificity are all present in the grower–auction transaction. Site specificity results from the auction being located in the production region, as growers themselves could bring the products to the auction. Physical asset specificity is present because the auction facilities are adjusted to the particular products of the growers. Temporal asset specificity is present because the products are perishable. Growers are the stakeholders that have most to win from investing in the auction facilities, and have most to lose when others control these assets.

Safeguarding from the growers' perspective is obtained through ownership, which entails control over the auction facilities and policies. Given the temporal asset specificity in transactions with perishable products, the cooperative auction puts much effort in improving the efficiency of logistic processes. As owners of the auction facilities, growers could not be held up in the logistic process. Registration of buyers also provided safeguards against opportunistic buyer behaviour. Safeguarding from the buyer perspective was hardly needed, because traders had not invested in specific assets. Where measurement problems could occur, they were prevented by the quality classification system of the auction and the reputation effect (social mechanism) inherent in frequent transactions. In the old days of the vegetable auction the products of each grower were sold separately, so buyers knew whose product they were buying. As transactions were repeated many times, reputation effects could work.

Coordination was obtained through making the market as transparent as possible and through standardization. Information costs were reduced because growers and buyers did not have to spend effort on studying supply and demand conditions; the auction clock immediately made prices known to everyone. The pooled interdependence that is characteristic of a collective sales organization required standardization of products and processes. Standardization of products was obtained through the quality classification system as well as the uniform packaging requirements. Standardization of work processes was obtained formally by private regulations as well as informally by routines. Coordination was also obtained through informal norms of behaviour for both growers and traders.

CHANGING TRANSACTIONS, NEW GOVERNANCE MECHANISMS

In the agri-food industry, a number of changes in market, policy and technology have taken place in recent years that have affected the choice of efficient governance structures. Hobbs and Young (2000) distinguish four relevant changes in the environment: shifting consumer preferences towards more variety, more convenience, higher quality and better safety guarantees; changes in legislation, such as in agricultural policies, environmental policies and food safety regulations; new technologies, such as ICT and biotechnology; and changes in market structure, particularly concentration in the food retail industry. These developments affect the characteristics of transactions with agri-food products, by increasing uncertainty, asset specificity and measurement problems as well as strengthening interdependencies (Royer and Rogers 1998; Hobbs 2003). No significant changes in frequency are found. We will now discuss the changes in the characteristics of transactions with fruits and vegetables, and their impact on safeguarding and coordination. Once again, the focus is on the first trading stage in the marketing channel, that is, the transaction between the grower and his direct customer.

Asset specificity: While Dutch fruits and vegetables were traditionally sold under a generic brand (Holland), nowadays each producer group or marketing organization seeks to sell under its own brand name. This can be a consumer brand or a business-to-business brand. Establishing a brand requires substantial investments in advertising and reputation building. The owner of the brand will seek safeguarding to protect its investments against opportunistic behaviour by any other firm that handles the branded products. Asset specificity is also present in the case of specific packaging stations. As more and more vegetables are sold prepacked, the packaging facilities are specific to the products of the grower: the packaging line has to be available when the products are harvested (temporal specificity) and it has to be adjusted to the variety and volume of the grower's product (physical asset specificity).

Measurement problems: Measuring performance in fresh-produce transactions has become more difficult because of the particular attributes these products may have, because more products are sold under a brand name, and because of innovation requiring effort of all chain partners. Food products increasingly have special attributes like environmentally friendly, animal-friendly, non-GMO or organic, which are difficult to measure at the product itself, leading to information asymmetry. From the trader's perspective, there is an adverse selection problem, from the grower's perspective there is a moral hazard problem. Related to the asset-specificity problem described above is the problem of measuring the performance of the trader in supporting the grower's brand. Difficult performance measurement is also present in the case of product innovations that require efforts of several chain partners to generate the full value at the final consumer. As the grower has probably invested most in this product innovation, he has most to lose from a lack of effort by the other chain partners.

Uncertainty / complexity: Uncertainty increases when competition increases, for instance as a result of a decrease in the number of buyers. Also the need for more product innovation increases uncertainty, because it is uncertain whether a new product will be a success with consumers. If product innovation requires special effort of all chain partners, transactions become more complex.

Interdependency: When using the auction channel, interdependencies are mainly of a pooled kind. When selling through other channels sequential interdependencies may increase, because of the following developments. First, product innovations may have a system character requiring coordinated effort of several chain partners (production, logistic providers, wholesaler, retailer). Second, quality control throughout the chain demands coordination among the activities of all chain partners. Third, improving logistic efficiencies demands coordination of all chain partners. Four producing customized products or products in customized packaging ties producers to their customers. The quintessence of sequential interdependence is that the transaction between parties A and B is interdependent with the transaction between B and C. In other words, the grower's production activities and the trader's marketing activities are interdependent.

In sum, the changes in the grower–customer transactions with fruits and vegetables give rise to increasing safeguarding costs due to asset specificity and difficulties in performance measurement, as well as to increasing coordination costs due to a shift from pooled to sequential interdependence. As a solution to the safeguarding problem we expect to see a shift in the distribution of property rights along the chain, while as solution to the coordination problem we expect to see a shift from standardization to direct supervision as the main coordination mechanism. Can we find these shifts in the Dutch fresh-produce chains?

TRANSFORMATION IN DUTCH FRESH-PRODUCE CHANNELS

In the Dutch fresh-produce industry there has been a shift from the dominant auction channel (model III) to other channel models, most notably a shift from III to IV. This shift entails that growers no longer sell to wholesalers (through the auction cooperative) but have vertically integrated into wholesaling. There have been two different models of growers vertically integrating downstream in the chain. First, the traditional auction cooperatives have transformed into marketing and wholesale cooperatives, and second, many new small marketing cooperatives have been set up by growers that terminate their membership of the auction cooperative.

The main example of growers vertically integrating downstream by taking over wholesale assets has been the establishment of The Greenery (Bijman 2002; Bijman and Hendrikse 2003). In 1996, nine out of 20 Dutch fruit and vegetable auctions merged into the new cooperative Voedingstuinbouw Nederland (VTN), and combined all assets and activities in one central marketing firm, called The Greenery BV. Cooperative VTN is the 100% shareholder of The Greenery. The goals of the new marketing cooperative were to reduce costs, increase scale of operation, add more value, enhance market orientation and improve coordination in the production and distribution chain (Bijman 2002). The next step in the transformation process was the 1998 acquisition of two fresh-produce wholesale companies. The Greenery is now by far the largest marketing cooperative for fresh produce in The Netherlands. With a turnover in 2003 of more than 1.5 billion euro, it sells about half of all vegetables produced in The Netherlands. The Greenery is a cooperative wholesale company that trades directly with major retailers in The Netherlands and abroad. It also imports fruits and vegetables, both exotic and those products that are out-of-season in The Netherlands. Its main marketing strategy is category management: supplying the full range of fruits and vegetables, year-round, to its retail clients. As part of its marketing strategy The Greenery is investing in establishing a brand name and a reputation of quality supplier. Thus, The Greenery is building up reputation assets. To protect and fully exploit these assets, it strives to have as much control over the distribution channel as possible.

At the same time that The Greenery was formed and transformed, many growers founded new marketing cooperatives. Bijman (2002) has found that 75 new grower associations and grower-owned marketing cooperatives have been established in the years 1995-2000. The goals of many of these new marketing cooperatives was to trade directly with retail customers and to build a reputation with large retailers or even consumers (for instance by establishing a brand name). Many of these new cooperatives focused on the high-quality part of the market, by selling high-quality products, customized products (mainly customized packaging) or exclusive products. As we have argued above, transactions with these products are characterized by safeguarding needs and by sequential coordination. Thus, gaining control over the main parts of the distribution chain, by vertically integrating into wholesale, is a solution to the safeguarding problem.

Thompson (1967) suggested that in case of sequential interdependence, co-ordination could be obtained by direct supervision. Direct supervision implies that one party has control over the interdependent transactions. In other words, the alignment between transactions A→B and B→C is obtained by giving one party the power to decide on the execution of both transactions. In the case of a grower-owned cooperative, control is divided between the growers and the management of the cooperative firm. Traditionally, in the cooperative auction control resided with the growers collectively. Auction management did not have much freedom to take decisions. In the new marketing cooperatives, the management of the cooperative firm has gained substantial decision-making power to regulate both quantity and quality of supplies. This shift in decision-making power applies to both strategic and operational decisions. Already with the establishment of The Greenery, a separation of responsibilities between the association of members (VTN) and the cooperative firm (The Greenery) was introduced, giving the management of the cooperative firm more freedom to take decisions. But also on the operational side there has been a shift. For instance, transportation from the grower to the cooperative facilities has always been the responsibility of the grower himself. In 2004, The Greenery has decided to transfer this function from the growers individually to The Greenery management. By having control over the logistics of the supply as well as delivery, The Greenery is better able to coordinate these two interdependent transactions.

Also in the newly established marketing cooperatives a substantial part of control over grower–cooperative transactions is transferred to the management, in order to obtain the coordination needed to deal with sequential interdependence. As the cooperative firm takes care of the packaging and marketing (under a brand name), growers have to comply with quality and quantity restrictions set by the management.

CONCLUSION

In this paper we have developed a model for the study of governance structure change. Where traditional research on governance structure choice works with a small set of (discrete) governance structures, we have shifted the level of analysis to the constituent governance mechanisms. In addition, we acknowledge that both formal and informal mechanisms may provide solutions for the safeguarding and coordination problems. When the characteristics of a transaction change, for instance because of changes in market structure, public policies or technology, transaction costs related to safeguarding and coordination may rise. By choosing a proper combination of mechanisms firms may obtain the efficient governance structure needed to support and regulate the transaction. We have presented a preliminary application of the model to governance structure changes in the Dutch fresh-produce industry. This industry has gone through substantial restructuring in recent years. Traditional governance structures are no longer popular, and various new governance structures have appeared.

Both propositions presented in this paper seem to be confirmed by the developments in the Dutch fresh-produce industry. Asset specificity has increased, mainly due to the introduction of brands and the establishment op specialized packaging stations. Measurement problems have increased due to specific quality attributes, product innovation and quality guarantees. As a solution, producers have vertically integrated downstream by taking over or setting up wholesale companies. Coordination problems have increased due to the shift from pooled to sequential interdependence. As a solution, coordination mechanisms have shifted from standardization to direct supervision. The latter has been materialized by giving the management of marketing cooperatives more authority.

Our preliminary assessment of the usefulness of the model has mainly been qualitative. The next step should be a further operationalizing of the various constructs developed in this paper, and to set up a quantitative study to analyse in more detail the relationship between changes in transaction characteristics and shifts in governance mechanisms.

REFERENCES

Barzel, Y., 1982. Measurement cost and the organization of markets. *Journal of Law and Economics,* 25 (1), 27-48.

Bijman, J., 2000. Cooperatives. *In:* Douw, L. and Post, J. eds. *Growing strong: the development of the Dutch agricultural sector; background and prospects.* LEI, The Hague, 127-133.

Bijman, J. and Hendrikse, G., 2003. Co-operatives in chains: institutional restructuring in the Dutch fruit and vegetables industry. *Journal on Chain and Network Science,* 3 (2), 95-107.

Bijman, W.J.J., 2002. *Essays on agricultural co-operatives: governance structure in fruit and vegetable chains.* Proefschrift Rotterdam [http://www.lei.wageningen-ur.nl/publicaties/PDF/2002/PS_xxx/PS_02_02.pdf]

Borgatti, S.P. and Foster, P.C., 2003. The network paradigm in organizational research: a review and typology. *Journal of Management,* 29 (6), 991-1013.

Cheung, S.N.S., 1983. The contractual nature of the firm. *Journal of Law and Economics,* 26 (1), 1-21.

Coase, R.H., 1937. The nature of the firm. *Economica,* 4 (16), 386-405. [http://people.bu.edu/vaguirre/courses/bu332/nature_firm.pdf]

Grandori, A., 1997. An organizational assessment of interfirm coordination modes. *Organization Studies,* 18 (6), 897-925.

Granovetter, M., 1992. Problems of explanation in economic sociology. *In:* Nohria, N. and Eccles, R.G. eds. *Networks and organizations: structure, form, and action.* Harvard Business School Press, Boston, 25-56.

Grossman, S.J. and Hart, O.D., 1986. The costs and benefits of ownership: a theory of vertical and lateral integration. *Journal of Political Economy,* 94 (4), 691-719.

Gulati, R. and Singh, H., 1998. The architecture of cooperation: managing coordination costs and appropriation concerns in strategic alliances. *Administrative Science Quarterly,* 43 (4), 781-814.

Hannan, M.T. and Freeman, J., 1989. *Organizational ecology.* Harvard University Press, Cambridge.

Hart, O. and Moore, J., 1990. Property rights and the nature of the firm. *Journal of Political Economy,* 98 (6), 1119-1158.

Hesterly, W.S., Liebeskind, J. and Zenger, T.R., 1990. Organizational economics: an impending revolution in organization theory. *Academy of Management Review,* 15 (3), 402-420.

Hobbs, J.E., 2003. Institutional adaptation in the agri-food sector. *In:* Van Huylenbroeck, G., Verbeke, W., Lauwers, L., et al. eds. *Importance of policies and institutions for agriculture: Liber Amicorum Prof.dr.ir. Laurent Martens.* Academia Press, Gent, 57-77.

Hobbs, J.E. and Young, L.M., 2000. Closer vertical co-ordination in agri-food supply chains. *Supply Chain Management,* 5 (3), 131-142.

Holmström, B. and Milgrom, P., 1991. Multitask principal agent analyses: incentive contracts, asset ownership, and job design. *Journal of Law, Economics & Organization,* 7, 24-52.

Holmström, B. and Milgrom, P., 1994. The firm as an incentive system. *American Economic Review,* 84 (4), 972-991.

Jones, C., Hesterly, W.S. and Borgatti, S.P., 1997. A general theory of network governance: exchange conditions and social mechanisms. *Academy of Management Review,* 22 (4), 911-945.

Kemmers, W.H., 1987. De groente- en fruitveilingen tot 1945. *In:* Plantenberg, P. ed. *100 jaar veilingen in de tuinbouw [1887-1987].* Centraal Bureau van de Tuinbouwveilingen in Nederland, 11-34.

Klein, B., Crawford, R.G. and Alchian, A.A., 1978. Vertical integration, appropriable rents, and the competitive contracting process. *Journal of Law and Economics,* 21 (2), 297-326.

Meulenberg, M.T.G., 1989. Horticultural auctions in The Netherlands: a transition form 'price discovery' institution to 'marketing' institution. *Journal of International Food & Agribusiness Marketing,* 1 (3/4), 139-165.

Milgrom, P. and Roberts, J., 1992. *Economics, organization and management.* Prentice-Hall Intern., Englewood Cliffs.

Mintzberg, H., 1979. *structuring of organizations: a synthesis of the research.* Prentice-Hall, Englewood Cliffs.

Plantenberg, P. (ed.) 1987. *100 jaar veilingen in de tuinbouw [1887-1987]*. Centraal Bureau van de Tuinbouwveilingen in Nederland.

Royer, J.S. and Rogers, R.T., 1998. *The industrialization of agriculture: vertical coordination in the U.S. food system*. Ashgate, Aldershot.

Thompson, J.D., 1967. *Organizations in action: social science bases of administrative theory*, New York.

Van Stuijvenberg, J.H., 1977. *De ontstaansgronden van de landbouwcooperatie in her-overweging*. Nationale Cooperatieve Raad, 's-Gravenhage.

Williamson, O.E., 1979. Transaction-cost economics: the governance of contractual relations. *Journal of Law and Economics,* 22 (2), 233-261.

Williamson, O.E., 1987. *The economic institutions of capitalism: firms, markets, relational contracting*. Free Press, New York.

Williamson, O.E., 1991. Comparative economic organization: the analysis of structural alternatives. *Administrative Science Quarterly,* 36, 269-296.

Williamson, O.E., 1999. Strategy research: governance and competence perspectives. *Strategic Management Journal,* 20 (12), 1087-1108.

CHAPTER 16

EFFECTIVE PARTNERSHIPS FOR AGRI-FOOD CHAINS

The impact of supply-chain partnerships on supplier performance in the UK fresh-produce industry

RACHEL DUFFY AND ANDREW FEARNE

Centre for Food Chain Research, Imperial College London, South Kensington campus, London SW7 2AZ. E-mail: a.fearne@imperial.ac.uk

Abstract. This paper presents a framework of buyer–supplier relationships used in an empirical study to investigate how the development of more collaborative relationships between UK retailers and fresh-produce suppliers affects the financial performance of suppliers in such relationships. Relationships between key partnership characteristics and performance are discussed and empirically tested. In addition, multivariate analysis is used to identify the dimensions of buyer–supplier relationships that make the greatest relative contribution to the explanation of the performance construct.

INTRODUCTION

Traditionally, inter-organizational linkages between firms have been arm's-length and often adversarial with individual firms seeking to achieve cost reductions or profit improvements at the expense of their buyers and/or suppliers. However, researchers, such as Lamming (1993) and Christopher (1998), state that successful companies recognize that the transfer of costs up and down the supply chain does not make firms any more competitive as ultimately all costs make their way back to the final marketplace. Instead firms that engage in co-operative long-term partnerships that help to improve the efficiency of the supply chain as a whole for the mutual benefit of all parties involved, are more likely to be successful.

The UK food industry has seen a concerted move in recent years towards fewer and more co-operative buyer–supplier relationships as retailers have attempted to gain more control over their supply chains. This has been done to ensure the integrity of their own label products, in terms of quality and safety issues, and to reduce supply-chain costs in an effort to increase their competitiveness in a highly competitive retailing environment (Fearne and Hughes 1999). These efforts have

C.J.M. Ondersteijn et al. (eds.), Quantifying the agri-food supply chain, 225-239.

been accelerated in recent years by the introduction of Efficient Consumer Response (ECR), which promotes the development of collaborative partnerships between retailers and suppliers (Mitchell 1997; Fiddis 1997).

ECR is based on the premise that many business practices and attitudes within the food industry are counter-productive, with firms seeking to maximize their own efficiency and profitability by passing problems and costs up or down the supply chain to their trading partners. Therefore, the fundamental aim of ECR is to apply a total systems view and encourage firms to work together to remove unnecessary costs from the supply chain and to add value to products by identifying and responding to consumer needs more effectively (Mitchell 1997; Fiddis 1997; Lamey 1996). Because ECR relies on a seamless flow of information throughout the supply chain, the benefit of ECR is dependent on a move away from traditional confrontational relationships to relationships based on co-operation and trust (Wood 1993; IGD ECR Methodology Approach Group 1996; Fiddis 1997; Mitchell 1997).

In the food industry partnerships have been promoted as offering mutual benefits to both retailers and suppliers. However the publicized benefits have referred primarily to the supply chain as a whole or to the retailer's operations (IGD ECR Methodology Approach Group 1996; Coopers and Lybrand 1996). In addition, anecdotal evidence that does exist in the food industry refers primarily to relationships between retailers and large branded manufacturers (Harlow 1994; Pearce 1997; Fiddis 1997; Mitchell 1997). As such there is virtually no evidence of the status or outcomes of partnership developments with suppliers in unbranded commodity sectors, such as fresh meat and fresh produce (e.g. fresh fruit, salads and vegetables).

Although moves towards more co-operative buyer–supplier relationships are evident in the food industry and much has been written about the creation of such partnerships in the extant literature, research that has investigated what these partnerships entail and that has examined the outcomes of these relationships is limited. This lack of research has been highlighted by researchers such as Stuart (1993), who notes, "empirical evidence of the benefits of partnerships is scant and primarily limited to the automotive industry". Similarly Heide and Stump (1995) state, "empirical evidence regarding performance is virtually non-existent and although recent evidence suggests that co-operative forms of buyer–supplier relationships are becoming increasingly common no study to date has formally examined their implications". More recently several other researchers have also commented on the lack of research regarding the performance outcomes of partnerships (e.g. Kalwani and Narayandas 1995; Sheth and Sharma 1997; Cannon and Homburg 2001).

To our knowledge there seems to be a complete lack of any UK research that attempts to quantify the outcomes of moves to greater collaboration between food retailers and their suppliers. These deficiencies in research suggest that an empirical investigation of the nature of buyer–supplier relationships and their implications for performance will make a useful contribution to both inter-organizational theory in general and our understanding of retailer–supplier partnerships in the UK food industry in particular.

Therefore, this research investigates how partnerships between UK food retailers and suppliers affect the financial performance of suppliers. The views of suppliers are of particular interest as most suppliers are developing their relationships in response to their retail customer's demands for increased service. As partnerships require suppliers to make substantial investments in terms of time and financial resources the costs of engaging in closer relationships, such as those promoted by the ECR initiative, could outweigh the benefits of doing so, particularly in commodity sectors which consist of many small and medium-sized businesses that typically operate on tight margins (Fearne and Hughes 1999).

THE CONCEPTUAL FRAMEWORK

The framework used to investigate buyer–supplier relationships was developed from two key disciplinary orientations in channel theory: the behavioural approach and the political economy paradigm.

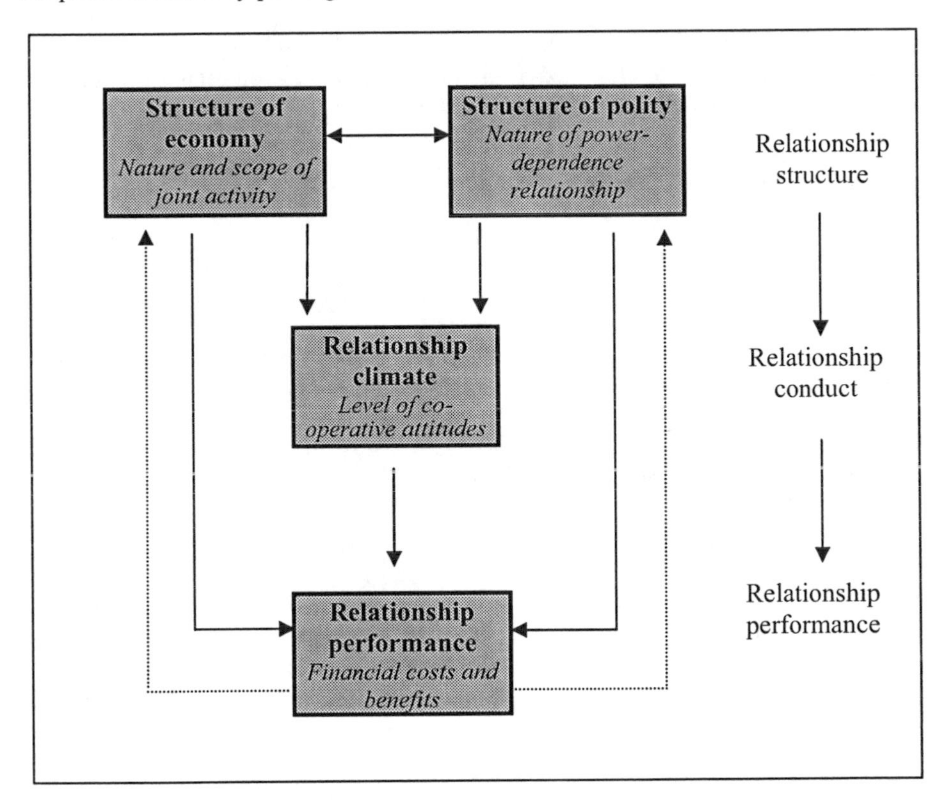

Figure 1. *Theoretical framework for investigating buyer–supplier relationships*

Building on the empirical work of Reve and Stern (1986) and the conceptual work of Robicheuax and Coleman (1994), who took a behavioural approach to the traditional structure–conduct–performance relationship, the premise of the model (Figure 1) is that the structural elements of a buyer–seller relationship, such as activities and information flows, measured in the internal economy, and the nature of the power–dependence relationship, measured in the internal polity, influence each other but also influence the dominant attitudes and sentiments in the relationship and the performance outcomes achieved. Each part of the framework is briefly discussed in the following sections. For a full discussion regarding the development of the model and its validation see Duffy and Fearne (2002).

Conceptualization of the structure of the economy

The internal economy is defined in terms of the types of activities, resources and information flows that are used to support and co-ordinate the operation of the buyer–supplier relationship (Arndt 1983; Reve and Stern 1986; Robicheaux and Coleman 1994; Cannon 1992). As such, the economy is conceptualized as existing on a continuum representing the more tangible and observable aspects of relationships. At one end, firms engage in low levels of joint activities and have low levels of operational integration and at the other they engage in high levels of joint activities and have high levels of operational integration.

Conceptualization of the structure of the internal polity

The internal political structure is conceptualized as the level and nature of interdependence that exists in a relationship (Kumar et al. 1995). Researchers state that a comprehensive view of interdependence must encompass both asymmetry and magnitude of interdependence as both describe the socio-political structure of a channel relationship (e.g. Kumar et al. 1995; Frazier and Antia 1995; Geyskens et al. 1996). Therefore an examination of the relationship polity directs attention to the level of total interdependence in the relationship (i.e. the sum of both firms' dependence) and the level of dependence asymmetry in the relationship (i.e. the difference in the firms' dependence scores).

Conceptualization of the climate

The climate examines the dominant attitudes and sentiments that exist in a buyer–supplier relationship (Reve and Stern 1986). In line with Reve and Stern (1986) researchers such as Stern and Reve (1980) and Skinner et al. (1992) suggest that conflict and co-operation are the two dominant sentiments that regulate exchange relationships.

Four theoretical constructs are used to capture whether the dominant attitudes and sentiments in relationships are co-operative or adversarial in nature. These are trust, commitment, relational norms and functional conflict resolution methods, which are constructs that indicate the presence of co-operative behaviour directed towards collective as opposed to individual goals (Dwyer et al. 1987; Anderson and Narus 1990; Heide and John 1992; Morgan and Hunt 1994; Anderson et al. 1994;

Cannon and Perreault 1997; Siguaw et al. 1998). Functional conflict resolution is measured instead of measuring the level of conflict in a relationship as researchers suggest that conflict is not always detrimental to a relationship (e.g. Robicheaux and El-Ansary 1976; Michie and Sibley 1979). Instead it is the manner in which partners resolve conflict that has implications for partnership success (Mohr and Spekman 1994).

Conceptualization of performance

The aim of this part of the framework is to examine the financial costs and benefits associated with different forms of buyer–supplier relationships. Because the focus of this study is concerned with the impact of partnerships on supplier performance, performance is viewed from the perspective of individual channel members. More specifically, the focus of performance concerns the supplier's overall view of the performance outcomes of a specific customer relationship. This view is taken because suppliers often have many customers. As such it would be difficult to isolate the impact of any individual relationship on overall performance at the firm level.

HYPOTHESIZED RELATIONSHIPS

Each of the three key dimensions of buyer–supplier relationships in Figure 1 are hypothesized as being key influences on performance. A brief review of the literature is given to support the hypothesized relationships between each of the constructs in the model and performance. It should be noted that each of these three dimensions was found to exist in higher amounts in relationships classified as partnerships, as opposed to arm's-length relationships (Duffy and Fearne 2002). Therefore the overriding hypothesis in the model is that partnerships improve performance.

The relationship between the internal polity and performance

In general, researchers suggest that the higher the level of interdependence in a relationship the better the implications for performance. For example, Mohr and Spekman (1994) and Gattorna and Walters (1996) suggest that the essence of successful partnerships is the extent of interdependence between the partners. Several other researchers also suggest that high bilateral dependence is related positively to performance (e.g. Anderson and Narus 1991; Buchanan 1992; Kumar et al. 1995; Lusch and Brown 1996).

With regard to the nature of asymmetry in the relationship, the dependence literature does not offer unambiguous performance implications. Instead two points of view exist regarding the relationship between dependence and performance; they are referred to as the opportunistic and benevolent perspectives (Gundlach and Cadotte 1994). The opportunistic perspective suggests that a dependence advantage will manifest exploitative tendencies. That is, the possession of more power (i.e. less dependence) will encourage action to gain a disproportionate share of resources from a less powerful partner (Beier and Stern 1969; Buchanan 1986; Noordewier et al. 1990; Gundlach and Cadotte 1994). On the other hand, the benevolent perspective emphasizes co-operative exchange as those with the greatest power are able to manipulate other members to act in ways that achieve greater positive results for the whole system (Beier and Stern 1969). Although there are a number of views on the relationship between the structure of interdependence and performance two hypotheses are posited from the literature.

H1(a). Suppliers in buyer–supplier relationships characterized by greater interdependence achieve higher levels of performance.

H1(b). Suppliers in buyer–supplier relationships characterized by greater dependence asymmetry achieve lower levels of performance.

The relationship between the internal economy and performance

Numerous articles routinely exhort both customer and supplier firms to seek collaborative relationships with each other as a way of improving performance. For example, Spekman (1988) states that in an attempt to gain greater competitive advantage, buyers are forging closer, more collaborative relationships with a smaller number of vendors. Similarly, Mohr and Spekman (1994) suggest that more successful partnerships exhibit higher levels of co-ordination than less successful partnerships, while Narus and Anderson (1987) suggest that successful working partnerships are marked by co-ordinated actions directed at mutual objectives across organizations. Kalwani and Narayandas (1995) also suggest that suppliers in long-term, closer relationships achieve a higher level of sales growth and profitability compared to supplier firms that used a transactional approach to servicing customers. Therefore the following hypothesis is posited:

H2. Suppliers engaging in buyer–supplier relationships characterized by higher levels of collaborative activity achieve higher levels of performance.

The relationship between climate and performance

The importance of variables such as trust and commitment are highlighted in the food industry initiative ECR, which emphasizes that the benefit of joint working between retailers and manufacturers would only be fully realized if there was a move away from confrontational relationships to relationships based on co-operation, openness and trust (Fiddis 1997; Mitchell 1997).

In the inter-organizational literature commitment and trust are frequently highlighted as key mediating variables that contribute to relationship success in terms of efficiency, productivity and effectiveness (e.g. Noordewier et al. 1990; Sherman 1992; Anderson and Weitz 1992; Morgan and Hunt 1994; Mohr and Spekman 1994; Gundlach et al. 1995; Siguaw et al. 1998). Researchers also suggest a positive relationship between the existence of relational norms and performance (Lusch and Brown 1996; Siguaw et al. 1998) and suggest that conflict can be productive for the relationship if disputes are resolved amicably (Anderson and Narus 1990; Morgan and Hunt 1994; Mohr and Spekman 1994). The hypothesized relationship between the climate and performance is posited as:

H3. Suppliers in buyer–supplier relationships characterized by higher levels of co-operative attitudes and sentiments achieve higher levels of performance.

METHODOLOGY

Data collection

Data were collected via a questionnaire sent to the managing directors of 337 fresh-produce suppliers who supplied food retailers or food service companies directly. The survey was administered in March 2001 and a total of 173 questionnaires were returned. Of these, 155 were deemed usable, resulting in a usable response rate of nearly 46 percent.

Suppliers were instructed to answer the questionnaire in relation to the customer with whom they had been doing business for the longest period of time. This was done to increase the likelihood that suppliers commented on a relationship that was properly formed and had established patterns of behaviour (Leuthesser 1997). The decision to specify the customer about whom suppliers should comment was made as Ellram and Hendrick (1995) suggest that if the decision is left to the supplier, the results will be biased in favour of high-performing relationships as given the choice, suppliers are most likely to pick their best customer relationships to discuss. It was believed that the selection of high-age group relationships would not bias the responses towards relationships with more partnership characteristics, as researchers such as Leuthesser (1997) and Blois (1996; 1997) state that the established patterns of behaviour in the relationship may or may not be relational in nature. This belief was supported by the results of an ANOVA analysis, which showed that there were no significant differences in any of the variables in the study when relationships were grouped according to age (Duffy and Fearne 2004).

Measures used

All theoretical constructs were measured using multiple item scales. The structure of the economy was measured using a 22-item scale designed to capture the task-related flows of activities, resources and information in a relationship. The structure of the polity was measured using parallel multiple-item scales; one to measure the supplier's view of its dependence on the chosen customer and the other to measure the supplier's view of their customer's dependence on their own firm. This method for measuring interdependence has been suggested and used in several previous studies (e.g. Buchanan 1992; Kumar et al. 1995; Lusch and Brown 1996; Frazier and Antia 1995; Geyskens et al. 1996).

To measure the dominant attitudes and sentiments in the exchange separate scales were developed to measure levels of trust, commitment, relational norms and functional conflict resolution methods. Trust was measured using a four-item scale that captured trust in a partner's honesty and trust in a partner's benevolence (Kumar et al. 1995). Commitment was measured using three items that captured the attitudinal and temporal components of commitment (Kumar et al. 1995; Wilson and Vlosky 1998). Relational norms were measured using eight items that measured four norms most frequently used to operationalize the construct of relationalism. These were solidarity, flexibility, mutuality and information exchange (e.g. Kaufmann and Stern 1988; Noordewier et al. 1990; Gundlach et al. 1995; Dant and Schul 1992; Heide and John 1992; Lusch and Brown 1996). Functional conflict resolution was measured using items that identify whether problems are resolved amicably or by resorting to threats using items drawn from previous studies (Salmond 1987; Gundlach et al. 1995; Morgan and Hunt 1994).

Finally, performance was measured using nine items that captured commonly cited benefits of partnerships. These items measured whether there had been a reduction in costs and a sharing of realized benefits (IGD ECR Methodology Approach Group 1996; Fiddis 1997; Mitchell 1997) and changes in sales and profits which Frazier et al. (1988) and Nielson (1997) suggest are the most important outcomes of partnerships. In addition, items were developed which captured the supplier's beliefs and expectations regarding the future prospects for the relationship and its future viability, as Woo and Willard (1983) and Stern and El-Ansary (1992) suggest that performance cannot be measured solely by past or current levels of sales and profitability, but should also include indicators of how the firm will do in the future.

Validation and modification of measures

Prior to the questionnaire being sent, all measures were reviewed by a panel of academic specialists in the area of fresh produce and buyer–supplier relationships and by a group of industry executives. This review resulted in minor changes to the wording of some questions.

After the data had been collected all measures were tested for their reliability and validity, using Cronbach's alpha and factor analysis. A factor analysis of each multiple item scale identified ten distinct and separate inter-organizational constructs that were used in all subsequent statistical analyses. These had alpha values ranging from 0.63 to 0.93, indicating that all scales were reliable (Duffy and Fearne 2002). These are listed in Table 1.

Table 1. Key dimensions of buyer–supplier relationships

Construct	Description
Economy	**Sum of economy factors 1 to 4**
Economy factor 1	Focus on supply-chain efficiency
Economy factor 2	Exclusive offerings
Economy factor 3	Scope and level of communication and joint activities
Economy factor 4	Involvement in decisions and planning
Polity	**Total interdependence and dependence asymmetry**
Total interdependence	Supplier dependence + customer dependence
Dependence asymmetry	Supplier dependence - customer dependence
Climate	**Sum of climate factors 1 to 3**
Climate factor 1	Trust and relational norms
Climate factor 2	Commitment
Climate factor 3	Functional conflict resolution methods
Performance	**Sum of performance factors 1 and 2**
Performance factor 1	Future growth
Performance factor 2	Current costs and sales

RESULTS

The data were analysed in three parts. Firstly, the hypotheses were tested using three regression models that estimated the separate influence of the economy, the polity and the climate on performance. Secondly a regression model was estimated that used all the theoretical constructs in their factor form to identify the joint predictive power of all of the variables in the framework. In addition, this model was used to determine the relative importance of each independent variable in the prediction of performance.

Hypothesis testing

Three regression models were estimated to test the hypotheses. Prior to conducting the regressions the data for each of the individual variables were checked to ensure that they met the general assumptions of normality, linearity and homoscedasticity that underlie multivariate analyses (Hair 1998). The results of these tests indicated that no serious violations of these assumptions existed in the data set (Duffy 2002).

The results of the separate regression models for the economy, climate and polity (models 1 to 3) are shown in Tables 2 and 3. Table 2 compares the three models in terms of the amount of variance that the construct accounted for as a whole, while

Table 3 shows the individual impact of the underlying dimensions of each construct on performance.

Table 2. *Total variance in performance accounted for by regression models 1 to 3*

Model	Variables *in the* model	R^2	Adj. R^2	*SEE*	F	Sig.
1	Economy: Factors 1-4	0.43	0.41	0.45	27.956	0.000**
2	Climate: Factors 1-3	0.64	0.64	0.35	90.906	0.000**
3	Polity: Total interdependence Dependence asymmetry	0.19	0.17	0.53	17.259	0.000**

Table 3. *Impact on performance of the variables in regression models 1 to 3*

Model	Variables in the model	β	T	Sig.
1	Constant		8.172	0.000**
	Economy factor 1	0.103	1.280	0.202
	Economy factor 2	0.014	0.220	0.826
	Economy factor 3	0.109	1.383	0.169
	Economy factor 4	0.513	6.232	0.000**
2	Constant		2.310	0.022*
	Climate factor 1	0.406	6.215	0.000**
	Climate factor 2	0.428	7.418	0.000**
	Climate Factor 3	0.144	2.535	0.012*
3	Constant		8.229	0.000**
	Total interdependence	0.360-0.277	4.884	0.000**
	Dependence asymmetry		-3.761	0.000**

**P<0.01, *P<0.05

Economy factor 1 = Focus on supply-chain efficiency, **Economy factor 2** = Exclusive offerings, **Economy factor 3** = Level and scope of communication and joint activities, **Economy factor 4** = Involvement in decisions and planning, **Climate factor 1** = Trust and relational norms, **Climate factor 2** = Commitment, **Climate factor 3** = Functional conflict resolution

Table 2 shows that on their own the factors that represent the economy construct accounted for 41.2 percent of the variance in the performance construct. Therefore, the results support the hypothesis that collaborative activity is positively related to performance. The results in Table 2 also indicate that of the four factors that represent the economy, factor 4 (involvement in decision making and planning) accounts for the greatest amount of variance in performance.

Model 2 shows that the factors that make up the climate significantly accounted for 63.7 percent of the variance in the performance variable (Table 2). Therefore hypothesis 3 is supported. An examination of the beta values in Table 3 shows that commitment was the best predictor of performance, followed by trust and relational norms, functional conflict resolution.

Finally, model 3 shows that the structure of interdependence significantly accounts for 17.4 percent of the variance in performance. Table 3 shows that dependence asymmetry has a significant negative relationship with performance, while total interdependence has a significant positive relationship with performance. Therefore the results support hypotheses 1(a) and 1(b). According to the beta values total interdependence explained more of the variance in performance than dependence asymmetry.

Identifying the key influences on performance

A regression model was also estimated using all the theoretical constructs in their factor form to identify which aspects of buyer–supplier relationships in the framework have the greatest influence on performance. Table 4 shows that together the nine variables significantly explained 64.2 percent of the variation in the performance variable. However, only four variables explained a significant amount of variation in the performance construct when all the variables in the framework were considered simultaneously (Table 5).

Table 4. *Total variance in performance accounted for*

Model	R^2	Adj R^2	SEE	F	Sig.
4	0.66	0.64	0.35	31.625	0.000

Table 5. *Impact on performance of individual variables*

Variables in the model	β	T	Sig.
Constant		1.366	0.174
Economy factors			
Supply-chain focus	-0.004	-0.063	0.950
Exclusive offerings	-0.061	-1.104	0.272
Frequency/scope: communication/joint activities	0.009	0.125	0.900
Involvement in decisions/ planning	0.157	1.988	0.049*
Climate factors			
Trust and relational norms	0.318	3.894	0.000**
Commitment	0.380	5.782	0.000**
Functional conflict resolution	0.129	2.269	0.025*
Polity factors			
Total interdependence	0.079	1.291	0.199
Dependence asymmetry	-0.007	-0.119	0.905

**P<0.01, *P<0.05

Table 5 shows that of the four significant predictors, two variables (trust and relational norms and commitment) were significant at the 0.01 level and two variables (involvement in decisions and planning and functional conflict resolution) were significant at the 0.05 level. Using the beta coefficients to compare the impact of each variable it can be seen that commitment accounted for the most variance in performance, followed by trust and relational norms, involvement in decisions and planning and finally the level of functional conflict resolution methods. Therefore, the results indicate that the sentiments and attitudes that underlie the exchange are more significant indicators of performance than the structural dimensions of relationships. The interpretation of the results could have been distorted by multicollinearity in the data set but, following recommendations by Gujarati (1992), a series of auxiliary regressions carried out on the set of independent variables showed that the level of multicollinearity in the set of independent variables was low (Duffy 2002).

CONCLUSIONS

The results provide support for the theory that partnerships can help a firm to improve its performance. This conclusion is based on the fact that each of the main partnership dimensions in the theoretical framework had a significant and positive relationship with performance. The exception was the relationship between dependence asymmetry and performance, which had a negative relationship as predicted and indicates that power imbalances have a detrimental effect on the sharing of partnership benefits. The results also showed that when considered together the variables in the framework significantly accounted for over 64 percent of the variation in performance. Although causality cannot be inferred from these results the research contributes to the body of knowledge that implies that partnerships can help a firm to improve its performance.

The results also showed that commitment and trust and relational norms had the greatest predictive ability in the multiple-regression analysis, followed by functional conflict resolution and involvement in decisions and planning. Therefore it is concluded from this research that while all three constructs in the framework are significant indicators of performance it is the softer, more intangible, aspects of buyer–supplier relationships that are the more reliable explanatory variables for performance.

This study contributes to inter-organizational theory as it provides empirical evidence of the performance implications of partnerships, which have been severely lacking in the literature. In particular, it has answered the calls of researchers such as Heide and John (1988), Heide and Stump (1995) and Kalwani and Narayandas (1995), who have stressed the need for empirical research that examines the outcomes of closer relationships and partnerships, particularly on the performance of supplier firms.

The finding that the attitudes and sentiments that exist in the buyer–supplier relationships have the greater relative influence on performance highlights the importance of the legally binding code of practice that has been introduced by the UK Competition Commission to govern relationships between retailers and their suppliers in the food industry (Competition Commission 2000). This code of practice was introduced after the Competition Commission found evidence that retailers had been abusing their position of power in the industry and engaging in a number of buying practices that adversely affected the competitiveness of suppliers. They found that this had resulted in a 'climate of apprehension' among many suppliers, many of whom would not identify the offending parties for fear of reprisals. This research suggests that by encouraging the development of co-operative attitudes, the code of practice will help to ensure that the benefits to suppliers increase and that they do not receive an unfair portion of the costs associated with exchange.

As this research is one of the first attempts to investigate the outcomes of different types of buyer–supplier relationships in the fresh-produce industry, it provides an important platform for further research in the area. In particular, as the inter-organizational variables in the theoretical framework accounted for a substantial and significant amount of the variation in the performance of suppliers, the framework developed in this study could be used as the basis for future empirical studies. However, further research is needed to gain a more complete understanding of the dynamics of successful customer relationships and the realities of forming collaborative partnerships in a low-margin commodity sector. To do this requires additional forms of research such as case studies, which would explore the inter-organizational variables in more detail. Ideally these should involve speaking to both the retailer and the supplier. Whilst this was not considered to be a viable option for empirical research in the food industry, it should prove to be more feasible using a case-study approach.

REFERENCES

Anderson, E. and Weitz, B., 1992. The use of pledges to build and sustain commitment in distribution channels. *Journal of Marketing Research,* 29, 18-34.

Anderson, J. and Narus, J., 1990. A model of distributor firm and manufacturer firm working partnerships. *Journal of Marketing,* 54, 42-58.

Anderson, J. and Narus, J., 1991. Partnering as a focused market strategy. *Californian Management Journal,* 33 (3), 95-113.

Anderson, J.C., Håkansson, H. and Johanson, J., 1994. Dyadic business relationships within a business network context. *Journal of Marketing,* 58 (4), 1-15.

Arndt, J., 1983. The political economy paradigm: foundation for theory building in marketing. *Journal of Marketing,* 47, 44-54.

Beier, F. and Stern, L., 1969. Power in the channel of distribution. *In:* Stern, L. ed. *Distribution channel: behavioural dimensions*. Houghton Mifflin Company, Boston, 92-116.

Blois, K., 1996. Relationship marketing in organisational markets: assessing its costs and benefits. *Journal of Strategic Marketing,* 4 (3), 181-191.

Blois, K., 1997. Are business to business relationships inherently unstable? *Journal of Marketing Management,* 13 (5), 367-382.

Buchanan, L., 1986. *The organisation of dyadic relationships indistribution channels: implications for strategy and performance*. Stanford University. Phd Thesis, School of Business.

Buchanan, L., 1992. Vertical trade relationships: the role of dependence and symmetry in attaining organisational goals. *Journal of Marketing Research,* 29, 65-75.

Cannon, J. and Perreault, W., 1997. *The nature of business relationships: working paper*. Department of Marketing, Colorado State University, Fort Collins.

Cannon, J.P., 1992. *A taxonomy of buyer-supplier relationships in business markets*. University of North Carolina, Chapel Hill.

Cannon, J.P. and Homburg, C., 2001. Buyer-supplier relationships and customer firm costs. *Journal of Marketing,* 65 (1), 29-43.

Christopher, M., 1998. *Logistics and supply chain management: strategies for reducing costs and improving services.* 2nd edn. Financial Times/Prentice Hall, London.

Competition Commission, 2000. *Supermarkets: a report on the supply of groceries from multiple stores in the United Kingdom: summary.* Competition Commission, London. [http://www.competition-commission.org.uk/rep_pub/reports/2000/446super.htm]

Coopers and Lybrand, 1996. *European value chain analysis study: a cornerstone for Efficient Consumer Response.* ECR Europe, Utrecht.

Dant, R.P. and Schul, P.L., 1992. Conflict resolution processes in contractual channels of distribution. *Journal of Marketing,* 56 (1), 38-54.

Duffy, R., 2002. *The impact of supply chain partnerships on supplier performance: a study of the UK fresh produce industry.* University of London, Wye. PhD Thesis, Imperial College at Wye

Duffy, R. and Fearne, A., 2002. *The development and empirical validation of a political economy model of buyer-supplier relationships in the UK food industry.* Centre for Food Chain Research, Imperial College at Wye, Wye. CFCR Discussion Paper Series no. 1.

Duffy, R. and Fearne, A., 2004. Buyer-supplier relationships: an investigation of moderating factors on the development of partnership characteristics and performance. *International Food and Agribusiness Management Review,* 7 (2), 1-25.

Dwyer, F.R., Schurr, P.H. and Oh, S., 1987. Developing buyer-seller relationships. *Journal of Marketing,* 51, 11-27.

Ellram, L.M. and Hendrick, T.E., 1995. Partnering characteristics: a dyadic perspective. *Journal of Business Logistics,* 16 (1), 41-64.

Fearne, A. and Hughes, D., 1999. Success factors in the fresh produce supply chain: insights from the UK. *Supply Chain Management,* 4 (3), 120-128.

Fiddis, C., 1997. *Manufacturer-retailer relationships in the food and drink industry: strategies and tactics in the battle for power.* Financial Times Retail and Consumer Publications, London.

Frazier, G.L. and Antia, K.D., 1995. Exchange relationships and interfirm power in channels of distribution. *Journal of the Academy of Marketing Science,* 23 (4), 321-326.

Frazier, G.L., Spekman, R.E. and O'Neal, C.R., 1988. Just in time exchange relationships in industrial markets. *Journal of Marketing,* 52, 52-67.

Gattorna, J.L. and Walters, D.W., 1996. *Managing the supply chain: a strategic perspective.* Macmillan Business, London.

Geyskens, I., Steenkamp, J.B.E.M., Scheer, L.K., et al. 1996. The effects of trust and interdependence on relationship commitment: a trans-Atlantic study. *International Journal of Research in Marketing,* 13, 303-317. [http://www.tilburguniversity.nl/faculties/feb/marketing/members/geyskens/research/1996.pdf]

Gujarati, D., 1992. *Essentials of econometrics.* McGraw-Hill, New York.

Gundlach, G. and Cadotte, E.R., 1994. Exchange interdependence and interfirm interaction: research in a simulated channel setting. *Journal of Marketing Research,* 31, 516-532.

Gundlach, G.T., Achrol, R.S. and Mentzer, J.T., 1995. The structure of commitment in exchange. *Journal of Marketing,* 59 (1), 78-92.

Hair, J.F., 1998. *Multivariate data analysis.* Rev. edn. Prentice-Hall International, London.

Harlow, P., 1994. Category management: a new era in FMCG buyer-supplier relationships. *Journal of Brand Management,* 2 (5), 289-295.

Heide, J.B. and John, G., 1988. The role of dependence balancing in safeguarding transaction-specific assets in conventional channels. *Journal of Marketing,* 52, 20-35.

Heide, J.B. and John, G., 1992. Do norms matter in marketing relationships? *Journal of Marketing,* 56, 32-44.

Heide, J.B. and Stump, R.L., 1995. Performance implications of buyer-supplier relationships in industrial markets: a transaction cost explanation. *Journal of Business Research,* 32 (1), 57-66.

IGD ECR Methodology Approach Group, 1996. *ECR process framework.* Institute of Grocery Development, Watford. [http://www.igd.com/cir.asp?cirid=68&Menuid=84]

Kalwani, M.U. and Narayandas, N., 1995. Long-term manufacturer-supplier relationships: do they pay off for supplier firms. *Journal of Marketing,* 59 (1), 1-16.

Kaufmann, P.J. and Stern, L.W., 1988. Relational exchange norms, perceptions of unfairness and retained hostility in commercial litigation. *Journal of Conflict Resolution,* 32 (3), 534-552.

Kumar, N., Scheer, L.K. and Steenkamp, J.B.E.M., 1995. The effects of perceived interdependence on dealer attitudes. *Journal of Marketing Research,* 32, 348-356.

Lamey, J., 1996. *Supply chain management: best practice and the impact of new partnerships.* FT Retail and Consumer Publishing, London. Financial Times Management Reports.

Lamming, R., 1993. *Beyond partnership: strategies for innovation and lean supply.* Prentice Hall, New York.

Leuthesser, L., 1997. Supplier relational behaviour: an empirical assessment. *Industrial Marketing Management,* 26 (3), 245-254.

Lusch, R.F. and Brown, J.R., 1996. Interdependency, contracting, and relational behavior in marketing channels. *Journal of Marketing,* 60 (4), 19-38.

Michie, D. and Sibley, S., 1979. Channel conflict, competition, and co-operation: theory and management. *In:* Lusch, R. and Zinszer, P. eds. *Contemporary issues in marketing channels.* The University of Oklahoma, Norman, 65-75.

Mitchell, A., 1997. *Efficient consumer response: a new paradigm for the European FMCG sector.* Financial Times Retail and Consumer Publishing, London.

Mohr, J. and Spekman, R., 1994. Characteristics of partnership success: partnership attributes, communication behaviour and conflict resolution techniques. *Strategic Management Journal,* 15 (2), 135-152.

Morgan, R. and Hunt, S., 1994. The commitment-trust theory of relationship marketing. *Journal of Marketing,* 58 (3), 20-38.

Narus, J. and Anderson, J., 1987. Distributor contributions to partnerships with manufacturers. *Business Horizons,* 30, 34-42.

Nielson, C.C., 1997. An empirical examination of the role of closeness in industrial buyer-seller relationships. *European Journal of Marketing,* 32 (5/6), 441-463.

Noordewier, T.G., John, G. and Nevin, J.R., 1990. Performance outcomes of purchasing arrangements in industrial buyer-vendor relationships. *Journal of Marketing,* 54, 80-93.

Pearce, T., 1997. Lessons learned from the Bird's Eye Wall's ECR initiative. *Supply Chain Management,* 2 (3), 99-106.

Reve, T. and Stern, L., 1986. The relationship between interorganisational form, transaction climate, and economic performance in vertical interfirm dyads. *In:* Pellegrini, L. and Reddy, S.K. eds. *Marketing channels: relationships and performance.* Lexington Books, Lexington, 75-102.

Robicheaux, R. and Coleman, J., 1994. The structure of marketing channel relationships. *Journal of the Academy of Marketing Science,* 22 (1), 38-51.

Robicheaux, R. and El-Ansary, A., 1976. A general model for understanding channel member behaviour. *Journal of Retailing,* 52 (4), 13-30.

Salmond, D., 1987. *When and why buyers and sellers collaborate: a resource dependence and efficiency view.* Department of Marketing Maryland. PhD Thesis, Department of Marketing Maryland

Sherman, S., 1992. Are strategic alliances working? *Fortune Magazine,* 126 (6), 77-78.

Sheth, J.N. and Sharma, A., 1997. Supplier relationships: emerging issues and challenges. *Industrial Marketing Management,* 26 (2), 91-100.

Siguaw, J.A., Simpson, P.M. and Baker, T.L., 1998. Effects of supplier market orientation on distributor market orientation and the channel relationship: the distribution perspective. *Journal of Marketing,* 62 (3), 99-111.

Skinner, S.J., Gassenheimer, J.B. and Kelley, S.W., 1992. Co-operation in supplier-dealer relations. *Journal of Retailing,* 68 (2), 174-193.

Spekman, R., 1988. Strategic supplier selection: understanding long-term buyer relationships. *Business Horizons,* 31 (4), 75-81.

Stern, L.W. and El-Ansary, A.I., 1992. *Marketing channels.* Prentice-Hall, Englewood Cliffs.

Stern, L.W. and Reve, T., 1980. Distribution channels as political economies: a framework for comparative analysis. *Journal of Marketing,* 44 (3), 52-64.

Stuart, F.I., 1993. Supplier partnerships: influencing factors and strategic benefits. *International Journal of Purchasing and Material Management,* 29 (4), 22-28.

Wilson, D.T. and Vlosky, R.P., 1998. Interorganizational information system technology and buyer-seller relationships. *Journal of Business and Industrial Marketing,* 13 (3), 215-234.

Woo, C. and Willard, G., 1983. *Performance representation in strategic management research: discussion and recommendations, presented at the 23rd Annual National Meetings of The Academy of Management, Dallas.* Academy of Management.

Wood, A., 1993. Efficient Consumer Response. *Logistics Information Management,* 6 (4), 28-40.

LIST OF PARTICIPANTS

Aramyan, A.L.	Business Economics Group, Wageningen University, Wageningen, The Netherlands
Bijman, W.J.J.	Department of Business Administration, Wageningen University, Wageningen, The Netherlands
Bremmers, H.J.	Business Administration Group, Wageningen University, Wageningen, The Netherlands
Bunte, F.H.J.	LEI, Wageningen University and Research Centre, The Hague, The Netherlands
De Vlieger, J.J.	LEI, Wageningen University and Research Centre, The Hague, The Netherlands
Fearne, A.P.	Centre for Food Chain Research, Imperial College London, United Kingdom
Franz, E.	Horticultural Production Chains Group, Wageningen University, Wageningen, The Netherlands
Gengenbach, M.F.G.	Environmental Economics and Natural Resources Group, Wageningen University, Wageningen, The Netherlands
Haverkamp, D.J.	Management Studies Group, Wageningen University, Wageningen, The Netherlands
Hobbs, E.	Department of Agricultural Economics, University of Saskatchewan, Saskatoon, Canada
Ingenbleek, P.T.M.	Marketing and Consumer Behaviour Group, Wageningen University, Wageningen, The Netherlands
Kleijnen, P.C.	Tilburg University, Tilburg, The Netherlands
Meuwissen, M.P.M.	Institute for Risk Management in Agriculture, Wageningen University and Research Centre, Wageningen, The Netherlands
Novoselova, A.	Business Economics Group, Wageningen University, Wageningen, The Netherlands
Ondersteijn, C.J.M.	Business Economics Group, Wageningen University, Wageningen, The Netherlands
Oude Lansink, A.G.J.M.	Business Economics Group, Wageningen University, Wageningen, The Netherlands
Schepers, H.E.	Agrotechnology and Food Innovations, Wageningen University and Research Centre, Wageningen, The Netherlands
Schiefer, G.W.	University of Bonn, Bonn, Germany
Soler, L.G.	INRA-LORIA, Ivry-sur-Seine, France

Sporleder, L.	Department of Agricultural, Environmental, and Development Economics, The Ohio State University, Columbus, USA
Surkov, I.	Business Economics Group, Wageningen University, Wageningen, The Netherlands
Valeeva, N.I.	Institute for Risk Management in Agriculture, Wageningen University and Research Centre, Wageningen, The Netherlands
Van der Vorst, J.G.A.J.	Logistics and Operations Research Group, Wageningen University, Wageningen, The Netherlands
Van Kooten, O.	Horticultural Production Chains Group, Wageningen University, Wageningen, The Netherlands
Van Plaggenhoef, W.	Management Studies Group, Wageningen University, Wageningen, The Netherlands
Verhallen, Th.M.M.	Tilburg University, Tilburg, The Netherlands
Vosough Ahmadi, B.	Business Economics Group, Wageningen University, Wageningen, The Netherlands
Wijnands, J.H.M.	Business Economics Group, Wageningen University, Wageningen, The Netherlands
Wubben, E.F.M.	Management Studies Group, Wageningen University, Wageningen, The Netherlands
Wysocki, F.	University of Florida, Gainesville, USA

Wageningen UR Frontis Series

1. A.G.J. Velthuis, L.J. Unnevehr, H. Hogeveen and R.B.M. Huirne (eds.): *New Approaches to Food-Safety Economics*. 2003
   ISBN 1-4020-1425-2; Pb: 1-4020-1426-0
2. W. Takken and T.W. Scott (eds.): *Ecological Aspects for Application of Genetically Modified Mosquitoes*. 2003
   ISBN 1-4020-1584-4; Pb: 1-4020-1585-2
3. M.A.J.S. van Boekel, A. Stein and A.H.C. van Bruggen (eds.): *Proceedings of the Frontis workshop on Bayesian Statistics and quality modelling*. 2003
   ISBN 1-4020-1916-5
4. R.H.G. Jongman (ed.): *The New Dimensions of the European Landscape*. 2004
   ISBN 1-4020-2909-8; Pb: 1-4020-2910-1
5. M.J.J.A.A. Korthals and R.J.Bogers (eds.): *Ethics for Life Scientists*. 2004
   ISBN 1-4020-3178-5; Pb: 1-4020-3179-3
6. R.A. Feddes, G.H.de Rooij and J.C. van Dam (eds.): *Unsaturated-zone Modeling*. Progress, challenges and applications. 2004 ISBN 1-4020-2919-5
7. J.H.H. Wesseler (ed.): *Environmental Costs and Benefits of Transgenic Crops*. 2005 ISBN 1-4020-3247-1; Pb: 1-4020-3248-X
8. R.S. Schrijver and G. Koch (eds.): *Avian Influenza*. Prevention and Control. 2005 ISBN 1-4020-3439-3; Pb: 1-4020-3440-7
9. W. Takken, P. Martens and R.J. Bogers (eds.): *Environmental Change and Malaria Risk*. Global and Local Implications. 2005
   ISBN 1-4020-3927-1; Pb: 1-4020-3928-X
10. L.J.W.J. Gilissen, H.J. Wichers, H.F.J. Savelkoul and R.J. Bogers, (eds.): *Allergy Matters*. New Approaches to Allergy Prevention and Management. 2006
    ISBN 1-4020-3895-X; Pb: 1-4020-3896-8
11. B.G.J. Knols and C. Louis (eds.): *Bridging Laboratory and Field Research for Genetic Control of Disease Vectors*. 2006
    ISBN 1-4020-3800-3; Pb: 1-4020-3799-6
12. B. Tress, G. Tress, G. Fry and P. Opdam (eds.): *From Landscape Research to Landscape Planning*. Aspects of Integration, Education and Application. 2006
    ISBN 1-4020-3979-4; Pb: 1-4020-3978-6
13. J. Hassink and M. van Dijk (eds.): *Farming for Health*. Green-Care Farming Across Europe and the United States of America. 2006
    ISBN 1-4020-4540-9; Pb: 1-4020-4541-7
14. R. Ruben, M. Slingerland and H. Nijhoff (eds.): *The Agro-Food Chains and Networks for Development*. 2006
    ISBN 1-4020-4592-1; Pb: 1-4020-4600-6
15. C.J.M. Ondersteijn, J.H.M. Wijnands, R.B.M. Huirne, and O. van Kooten (eds.): *Quantifying The Agri-Food Supply Chain*. 2006
    ISBN 1-4020-4692-8; Pb: 1-4020-4693-6

Printed in the United States
88786LV00001B/204/A

9 781402 046933